A SHORT COURSE IN AUTOMORPHIC FUNCTIONS

Joseph Lehner

Dover Publications, Inc.
Mineola, New York

Copyright

Bibliographical Note

This Dover edition, first published in 2014, is an unabridged republication of the work originally published in 1966 by Holt, Rinehart and Winston, Inc., New York, as part of the "Athena Series: Selected Topics in Mathematics." This edition is published by special arrangement with Cengage Learning, Inc., Belmont, California.

Library of Congress Cataloging-in-Publication Data

Lehner, J. (Joseph), 1912–
A short course in automorphic functions / Joseph Lehner. — Dover edition.
p. cm. — (Athena series. Selected topics in mathematics)
Originally published: New York : Holt, Rinehart and Winston, 1966.
Includes bibliographical references and index.
ISBN-13: 978-0-486-78974-3
ISBN-10: 0-486-78974-8
1. Automorphic functions. I. Title.

QA351.L42 2014
515'.9—dc23

2014025812

Manufactured in the United States by Courier Corporation
78974801 2014
www.doverpublications.com

Preface

This book was written from lecture notes of a one-semester course in the theory of discontinuous groups and automorphic functions, a subject in which interest is increasing at the present time. It will be useful also to mathematicians wishing to gain some acquaintance with the subject.

The book presupposes only the usual first courses in complex analysis, topology, and algebra. (The Lebesgue integral is used in only two places.) It will therefore be accessible to many first-year graduate students.

Chapter 1 elaborates the theory of discontinuous groups by the classical method of Poincaré, which utilizes his model of the hyperbolic plane. The needed hyperbolic geometry is developed in the text.

In Chapter 2 we develop automorphic functions and forms via the Poincaré series. The formulas for divisors of a function and form are proved and their consequences analyzed. An invariant scalar product is introduced, enabling us to discuss vector spaces of automorphic forms.

Chapter 3 is devoted to the connection between automorphic function theory and Riemann surface theory. This side of the subject has been treated rather lightly in text books on Riemann surfaces. The chapter closes with some applications of the Riemann-Roch theorem.

The exercises in the text range from routine verifications to significant theorems. The more difficult ones are accompanied by indication of proof. There are notes at the end of each chapter describing further results and extensions.

The book includes a glossary, with references to the definitions of the terms. A list of standard text books appears at the end of the book.

Our grateful thanks are due to a number of colleagues who attended the lectures, and most especially to Professor A. F. Beardon, who read the entire manuscript with great care. The typing of the handwritten draft was done most competently by my wife, and she also prepared the glossary and index.

JOSEPH LEHNER

College Park, Maryland
December, 1965

Contents

A SHORT COURSE IN
AUTOMORPHIC FUNCTIONS

[I]

Discontinuous Groups

An analytic function f is called automorphic with respect to a group Γ of transformations of the plane if f takes the same value at points that are equivalent under Γ. That is,

$$f \circ V(z) = f(z) \tag{1}$$

for each $V \in \Gamma$ and each $z \in D$, the domain of f. If we want to have nonconstant functions f, we must assume there are only finitely many equivalents of z lying in any compact part of D. This property of Γ is known as *discontinuity*.

The most important domain for f from the standpoint of applications is the upper half-plane† H. Now from (1) with f analytic in H we deduce that V is analytic in H and maps H into itself. It is natural to require that V be one-to-one in order that V^{-1} should be single-valued. Hence V is a linear-fractional transformation. The group Γ will therefore be a group of linear-fractional transformations, or as we shall call them, linear transformations.

The present chapter is devoted to a study of discontinuous groups of linear transformations.

1. Linear Transformations

1A. A *linear transformation* is a nonconstant rational function of degree 1; that is, a function

$$w = \frac{\alpha z + \beta}{\gamma z + \delta}, \qquad \alpha\delta - \beta\gamma \neq 0 \tag{2}$$

where $\alpha, \beta, \gamma, \delta$ are complex numbers and z is a complex variable. The function w is defined on all of the complex sphere Z except $z = -\delta/\gamma$ and $z = \infty$. With the usual convention that $u/0 = \infty$ for $u \neq 0$, we have

$$w\left(-\frac{\delta}{\gamma}\right) = \infty,$$

and we obtain $w(\infty)$ by continuity:

$$w(\infty) = \lim_{z\to\infty} w(z) = \lim_{z\to\infty} \frac{\alpha + \beta/z}{\gamma + \delta/z} = \frac{\alpha}{\gamma}.$$

† One could just as well use the unit disk, which is conformally equivalent to H.

In particular, $-\delta/\gamma$ will be ∞ if and only if $\gamma = 0$, and in that case $\alpha/\gamma = \infty$: the infinite points of the two planes then correspond under the mapping w.

As a rational function, w is regular in Z except for a simple pole at $z = -\delta/\gamma$. Suppose $\gamma = 0$. Then necessarily $\delta \neq 0$, $\alpha \neq 0$ (because of $\alpha\delta - \beta\gamma \neq 0$) and

$$w = \frac{\alpha}{\delta} z + \frac{\beta}{\delta}.$$

Hence $dw/dz = \alpha/\delta \neq 0$ and w is conformal at every finite z. At infinity we must use the uniformizing variables $z' = 1/z$, $w' = 1/w$; then we find that

$$w' = \frac{\delta z'}{\alpha} + \cdots,$$

and w' is conformal at $z' = 0$, which by definition means that w is conformal at $z = \infty$.

If $\gamma \neq 0$, we have

$$\frac{dw}{dz} = \Delta \cdot (\gamma z + \delta)^{-2}, \qquad \Delta = \alpha\delta - \beta\gamma;$$

hence $dw/dz \neq 0$ and w is conformal except possibly at $z = -\delta/\gamma$, $z = \infty$. At $z = -\delta/\gamma$ we must use the variables $z' = z + \delta/\gamma$, $w' = 1/w$:

$$w = -\frac{\Delta}{\gamma^2}\frac{1}{z + \delta/\gamma} + \frac{\alpha}{\gamma},$$

$$w' = -\gamma^2\Delta^{-1}z' + Bz'^2 + \cdots,$$

so that $(dw'/dz')_{z'=0} \neq 0$. At $z = \infty$ the correct variables are $z' = 1/z$, $w' = w$ and we get

$$w' = \frac{\alpha}{\gamma} - \frac{\Delta}{\gamma^2} z' + \cdots,$$

yielding the same conclusion.

Solving (2) for z we get

$$z = \frac{\delta w - \beta}{-\gamma w + \alpha}, \qquad \delta\alpha - \beta\gamma \neq 0$$

which is also a linear transformation and so is defined on the extended w-plane. The mapping $z \to w$ is therefore onto and hence one-to-one, and we may write $z = w^{-1}$.

Putting these results together we can assert:

THEOREM. The linear transformation (2) is a one-to-one conformal mapping of all of Z on itself.

For this reason a linear transformation is also called a *conformal automorphism* of Z.[1]

If two linear transformations are equal,

$$\frac{\alpha z + \beta}{\gamma z + \delta} = \frac{\alpha' z + \beta'}{\gamma' z + \delta'}, \qquad \alpha\delta - \beta\gamma \neq 0, \qquad \alpha'\delta' - \beta'\gamma' \neq 0$$

for all z, then each function has the same zero and the same pole, and this gives[†]

$$\frac{\alpha}{\alpha'} = \frac{\beta}{\beta'} = \lambda, \qquad \frac{\gamma}{\gamma'} = \frac{\delta}{\delta'} = \mu, \qquad \lambda \neq 0, \qquad \mu \neq 0$$

from which

$$\frac{\alpha z + \beta}{\gamma z + \delta} = \frac{\lambda}{\mu} \frac{\alpha' z + \beta'}{\gamma' z + \delta'}.$$

It follows that $\lambda = \mu$: two equal transformations have proportional coefficients, and the converse is obviously true. But multiplying the coefficients of a transformation by λ multiplies its determinant by λ^2. This permits us to normalize the transformation by requiring its determinant $\alpha\delta - \beta\gamma$ to be 1. We shall do this from now on: *a linear transformation is a function*

$$w = \frac{az + b}{cz + d}, \qquad ad - bc = 1 \tag{3}$$

where a, b, c, d are complex numbers. We usually write (3) in the form $w = Tz$.

Occasionally we shall make use of functions of the form (2)—that is, linear fractional mappings with nonzero determinant not necessarily equal to 1. Such functions will be called *bilinear mappings.*

Let $\bar{\Omega}_C$ denote the set of all linear transformations. We can introduce a multiplication into $\bar{\Omega}_C$ by defining, for $w = Tz$, $u = Sw$,

$$u = ST(z) = S(Tz).$$

That is, the product of two linear transformations is simply their composition regarded as mappings.[‡] It is seen that STz is a linear transformation if we recall that the determinant of STz is the product of the determinants of Sw and Tz and so is 1. We have observed that each transformation has an inverse that is again a linear transformation, and the transformation $Iz = z = (1 \cdot z + 0)/(0 \cdot z + 1)$ serves as an identity. Since the associative law is immediately verified, we conclude that *$\bar{\Omega}_C$ is a group.*

As we saw above, a linear transformation Tz is equal to each of

[†] The cases in which one (or possibly two) of the coefficients vanish must be treated separately.

[‡] Products of linear transformations are therefore read from right to left.

the transformations $(\rho T)z$, $\rho \neq 0$, and to no others. Since $\det T = \det(\rho T) = \rho^2 \det T = 1$, we must have $\rho = \pm 1$. A linear transformation, therefore, determines its coefficients up to a factor of ± 1. It is then reasonable to associate to the linear transformation (3) the two *matrices* $\pm T$, where

$$T = \begin{pmatrix} a & b \\ c & d \end{pmatrix}, \qquad ad - bc = 1. \tag{4}$$

Moreover, the product of two linear transformations corresponds to the product of their matrices, as one calculates easily.

This observation suggests that we should identify $\bar{\Omega}_C$ with the group of all 2×2 unimodular matrices with complex entries, a group we shall denote by Ω_C or $SL(2, C)$. Indeed, if we define

$$f : T \to Tz,$$

then f is a mapping from Ω_C onto $\bar{\Omega}_C$, and

$$f(ST) = f(S)f(T);$$

that is, f is a homomorphism onto. Its kernel certainly includes $\pm I$, where[†] $I = (1\ 0 \mid 0\ 1)$, and, by what has been said, can include nothing else. Hence

$$\Omega_C/\{I, -I\} \cong \bar{\Omega}_C.$$

The identification of Ω_C with $\bar{\Omega}_C$ is permissible provided we identify each $T \in \Omega_C$ with $-T$.

Let $T \in \Omega_C$ and let S be a bilinear mapping. Then $STS^{-1} \in \Omega_C$. In general, if T has a certain property with respect to a set A, then STS^{-1} has the same property with respect to the set $S(A)$. This enables us to assume, for example, that a certain special point is ∞, etcetera. The transition from T to STS^{-1} is merely a change of coordinates in the plane.

One of the best-known properties of a linear transformation is that it maps a circle or straight line onto a circle or straight line. This can be seen as follows. The equation

$$Az\bar{z} + Bz + \bar{B}\bar{z} + C = 0, \tag{5}$$

with A, C real, includes all circles ($A \neq 0$) and all straight lines ($A = 0$). Writing $z = (aw + b)/(cw + d)$, $ad - bc = 1$, and substituting, we find

$$A'w\bar{w} + B'w + \bar{B}'\bar{w} + C' = 0,$$

† Throughout the book we write $(a\ b \mid c\ d)$ for the matrix

$$\begin{pmatrix} a & b \\ c & d \end{pmatrix}.$$

where

$$A' = Aa\bar{a} + Ba\bar{c} + \bar{B}\bar{a}c + Cc\bar{c}$$

$$C' = Ab\bar{b} + Bb\bar{d} + \bar{B}\bar{b}d + Cd\bar{d}.$$

Since A' and C' are obviously real, this is also the equation of a circle or straight line.[2]

1B. Linear transformations are classified by their fixed points (solutions of $Tz = z$). When $c \neq 0$ in $T = (a\, b \mid c\, d)$, the fixed points are given by the equation

$$cz^2 + (d - a)z - b = 0. \tag{6}$$

There are two finite solutions $z = \xi_1, \xi_2$ of this equation:

$$\xi_1 = \frac{a - d + (\chi^2 - 4)^{1/2}}{2c}, \qquad \xi_2 = \frac{a - d - (\chi^2 - 4)^{1/2}}{2c} \tag{7}$$

where $\chi = a + d$ is the *trace*† of T. (Note use of $ad - bc = 1$ in this calculation.) The two fixed points are coincident if and only if $\chi = \pm 2$; then the transformation is called *parabolic.*

LEMMA 1. The trace is invariant under $T \rightarrow ATA^{-1}$, where A is a nonsingular matrix.

This follows at once from $\chi(AB) = \chi(BA)$.

When $c = 0$, we have $T\infty = \infty$; that is, ∞ is always one fixed point. T is then

$$Tz = \frac{az + b}{d}, \qquad ad = 1$$

and $b/(d - a)$ is the other fixed point if $d \neq a$. When $d = a$, T reduces to

$$Tz = z \pm b,$$

a translation, and we arbitrarily call ∞ the second fixed point. The two fixed points are coincident. Thus T is parabolic and we observe that $\chi = 2$, as it should be. *A parabolic transformation with fixed point* ∞ *is a translation, and conversely.* Finally, if $b = 0$, T is the identity.

There are never more than two fixed points unless T is the identity. A corollary of this remark is the following: *if three distinct points* z_1, z_2, z_3 *have the same images under* S *and* T, *then* $S = T$. For $S^{-1}T$ fixes each z_i and so is the identity.

† Because of the relation between matrices and linear transformations, the trace is indefinite up to a factor of ± 1.

Let us first suppose ξ_1, ξ_2 are finite and distinct ($c \neq 0, \chi^2 \neq 4$). Set up the transformation $W = W(z)$,

$$\frac{W - \xi_1}{W - \xi_2} = \kappa \frac{z - \xi_1}{z - \xi_2}, \qquad \kappa \neq 0, 1.$$

W has fixed points ξ_1 and ξ_2, and if $W(\infty) = w(\infty)$ where $w = Tz$, then W will be the same transformation as w. Since $w(\infty) = a/c$, this gives $\kappa = (a - c\xi_1)/(a - c\xi_2)$. As thus calculated, κ is always finite, for $\xi_2 = a/c$ combined with (6) would yield $ad - bc = 0$. We call

$$\frac{w - \xi_1}{w - \xi_2} = \kappa \frac{z - \xi_1}{z - \xi_2}, \qquad \kappa = \frac{a - c\xi_1}{a - c\xi_2} \tag{8}$$

the normal form of $w = Tz$; κ is called the *multiplier* of T. But $1/\kappa$ may also be regarded as the multiplier of T since we can write the transformation in the form

$$\frac{w - \xi_2}{w - \xi_1} = \frac{1}{\kappa} \frac{z - \xi_2}{z - \xi_1}.$$

In other words, whether we regard κ or $1/\kappa$ as the multiplier depends on which fixed point is labeled ξ_1 and which ξ_2. As a way out of the ambiguity let us define the multiplier to be the *pair* $(\kappa, 1/\kappa)$.

Since $\kappa + \kappa^{-1}$ is a symmetric function of ξ_1, ξ_2, it must be a rational function of the coefficients of (6). A calculation using $\xi_1 + \xi_2 = (a - d)/c$, $\xi_1\xi_2 = -b/c$ yields

$$\kappa + \kappa^{-1} = \chi^2 - 2. \tag{9}$$

LEMMA 2. The multiplier is invariant under $T \to ATA^{-1}$, where A is a bilinear mapping.

Indeed, (9) and Lemma 1 give $\kappa + \kappa^{-1} = \kappa' + \kappa'^{-1}$, where $\kappa' = \kappa(ATA^{-1})$. Hence $\kappa = \kappa'$ or $\kappa\kappa' = 1$. Thus the multiplier of T' is the pair $(\kappa', 1/\kappa')$, which is the same as the pair $(\kappa, 1/\kappa)$.

When $c = 0$ but $\chi^2 \neq 4$, there is but one finite and one infinite fixed point, and the normal form of T is

$$w - \xi_2 = \kappa(z - \xi_2), \qquad \kappa = \frac{a}{d}, \qquad \xi_2 = \frac{b}{d - a}. \tag{8a}$$

We find that (9) is still valid (remember $ad = 1$). Hence Lemma 2 holds.

Next, assume $\xi_1 = \xi_2 \neq \infty$ (that is, $c \neq 0, \chi = \pm 2$). The transformation

$$\frac{1}{w - \xi_1} = \frac{1}{z - \xi_1} \pm c, \qquad \xi_1 = \frac{a - d}{2c}, \tag{10}$$

where the sign before c is the sign of χ, has the unique fixed point ξ_1; moreover

$w(\infty) = T(\infty)$. Therefore (10) is the normal form for a parabolic transformation with finite fixed point. When $\xi_1 = \xi_2 = \infty$, the normal form is

$$w = z + b. \tag{10a}$$

We arbitrarily define $\kappa = 1$ in these cases in order to satisfy (9) and verify Lemma 2.

Lemma 2 has now been proved for all linear transformations.

On the assumption that T is not parabolic, κ can have any value other than 0 or 1. Writing $\kappa = \rho e^{i\theta}$, $\rho > 0$, $0 \leq \theta < 2\pi$, we classify as follows:

$\rho = 1$, $\theta \neq 0$, T is called *elliptic*

$\rho \neq 1$, $\theta = 0$, T is called *hyperbolic*

$\rho \neq 1$, $\theta \neq 0$, T is called *loxodromic.*

When T is parabolic, $\kappa = 1$ and $\chi = \pm 2$. In the nonparabolic case we use (9) and deduce the following;

THEOREM. A necessary and sufficient condition that T be elliptic, hyperbolic, or parabolic is that χ be real and $|\chi| < 2$, $|\chi| > 2$, or $|\chi| = 2$, respectively. A necessary and sufficient condition that T be loxodromic is that χ be nonreal. The transformations T and ATA^{-1}, A a bilinear mapping, are simultaneously elliptic, hyperbolic, etcetera.

In this book we shall have no use for loxodromic transformations.

Exercise 1. A linear transformation T is *periodic of order n* if and only if $n > 1$ is the smallest integer for which $T^n = I$. Prove that T is of order n if and only if T is elliptic and its multiplier is a primitive nth root of unity. Show that the trace of T is $2 \cos \pi l/n$, $(l, n) = 1$, where (l, n) is the greatest common divisor of the integers l and n.

Exercise 2. Let T be nonelliptic. If for some z and some sequence $(n) \to \infty$ we have $T^n z \to z_0$, then z_0 is a fixed point of T. Is the result true for elliptic T?

1C. In this book we shall be concerned almost entirely with linear transformations that map the upper half-plane H on itself. Such transformations correspond to matrices with real entries. Indeed, if T is in this class, it maps the real axis E on itself (by continuity of T and T^{-1}). Now $z \in \text{E}$ only if $z = \bar{z}$; hence $Tz = \bar{T}\bar{z} = \bar{T}z$ and so $T = \bar{T}$. Since the mapping T determines its coefficients up to a factor of ± 1, we either have $a = \bar{a}, \cdots, d = \bar{d}$ or $a = -\bar{a}, \cdots, d = -\bar{d}$; that is, either a, b, c, d are all real or all are pure imaginary. But in the latter case we would have

$$\text{Im } T(i) = \frac{\text{Im}\{(ai + b)(-\bar{c}i + \bar{d})\}}{|ci + d|^2} = \frac{bc - ad}{|ci + d|^2} < 0$$

and $T(i) \notin$ H. Hence a necessary condition that T preserve H is that it have real coefficients. This condition is clearly sufficient, since a linear transformation with real coefficients obviously preserves E, and because of the determinant condition[†] it maps H into H.

Call Ω_R the subgroup of Ω_C consisting of matrices with real entries. We have proved

THEOREM 1. $T \in \Omega_R$ if and only if T preserves H.

An element of Ω_R will be called a *real transformation* or a *real matrix*. It is clear that a real transformation preserves the real axis as well as the lower half-plane. A real transformation is never loxodromic; it is elliptic, hyperbolic, or parabolic according as $\chi = a + d$ is, in absolute value, less than 2, greater than 2, or equal to 2. An elliptic transformation has two conjugate nonreal fixed points; a hyperbolic transformation has two real fixed points; a parabolic transformation has one real fixed point.

A fixed circle of T is a circle or straight line that is mapped on itself by T. The easiest way to discuss the fixed circles is to make a nonsingular transformation $z \to z'$ of the plane that carries the fixed points of T to 0 and ∞ or, in the case of parabolic T, carries the fixed point to ∞. In the first case T becomes $T'z' = \kappa z'$ with the same κ as in T (see 1B, Lemma 2). When T is elliptic, $\kappa = e^{i\theta}$, and T' is a rotation about the origin; the fixed circles are circles with center at the origin and each fixed circle is orthogonal to the family of rays through the origin—that is, orthogonal to the family of lines joining the fixed points. When T is hyperbolic, $\kappa > 0$, and T' is a stretching from the origin; the fixed circles are rays through the origin. When T is parabolic, $T'z' = z' + b$, a translation; the fixed circles are horizontal lines. In each case T preserves the interior of the fixed circle, or one side of the straight line if the fixed circle is a line. Applying the inverse mapping $z' \to z$ we now have

THEOREM 2. Each fixed circle of an elliptic T contains one and only one of the fixed points of T and is orthogonal to the family of circles joining the fixed points of T. Each fixed circle of a hyperbolic T passes through the fixed points of T. Each fixed circle of a parabolic T is a circle tangent to E at the fixed point of T. In every case T preserves the interior of a fixed circle.

In particular, E is a fixed circle of every $T \in \Omega_R$.

Later we shall interpret real linear transformations as noneuclidean rigid motions (4B).

[†] In connection with real transformations T a useful formula is

$$y' = \frac{y}{| cz + d |} = \frac{y}{\{(cx + d)^2 + c^2y^2\}},$$

where $z = x + iy, \quad Tz = x' + iy'$.

1D. Occasionally we shall have to do with linear transformations that map the open unit disk U on itself. Such transformations also preserve Q, the boundary of U. Call the subgroup of these transformations Ω_Q.

If A is a bilinear mapping of H on U, then $\Omega_Q = A\,\Omega_R A^{-1}$, and we can find the properties of Ω_Q from those of Ω_R. For example, take

$$A = \begin{pmatrix} 1 & -i \\ 1 & i \end{pmatrix}, \qquad A^{-1} = \frac{1}{2}\begin{pmatrix} 1 & 1 \\ i & -i \end{pmatrix};$$

then for $T = (a\, b \mid c\, d) \in \Omega_R$,

$$ATA^{-1} = \frac{1}{2}\begin{pmatrix} a + d + i(b - c) & a - d - i(b + c) \\ a - d + i(b + c) & a + d - i(b - c) \end{pmatrix},$$

and this is of the form

$$V = \begin{pmatrix} \alpha & \bar{\gamma} \\ \gamma & \bar{\alpha} \end{pmatrix}, \qquad \alpha\bar{\alpha} - \gamma\bar{\gamma} = 1. \tag{11}$$

But V obviously preserves U, for with $\zeta \in$ Q we have

$$|V\zeta| = \left|\frac{1}{\zeta} \cdot \frac{\alpha\zeta + \bar{\gamma}}{\gamma + \bar{\alpha}\bar{\zeta}}\right| = 1,$$

and $|V(0)| = |\bar{\gamma}/\bar{\alpha}| < 1$ from the determinant condition. Hence Ω_Q consists of the matrices (11).

Since $\alpha + \bar{\alpha}$ is real, there are no loxodromic elements in Ω_Q. An elliptic element has fixed points inverse to Q (that is, $\xi_1 = 1/\bar{\xi}_2$) and not on Q; a hyperbolic or a parabolic element has fixed points (point) on Q.

1E. THEOREM. Two real linear transformations, neither the identity, are commutative if and only if they have the same set of fixed points.

The "if" part is easy, for the transformations V_1, V_2 are both parabolic or both nonparabolic. In the first case we may assume, by transforming the plane, that the common fixed point is ∞. Then V_1, V_2 become translations and obviously commute.† In the second case we take the common fixed points to be 0 and ∞; then $V_i z = \kappa_i z$, $i = 1, 2$, and the conclusion is obvious.

A proof of the converse may be given by calculating the commutator of V_1 and V_2 in the various cases. Here is another proof. Suppose V_1 and V_2 commute and V_1 is parabolic with fixed point α. Then $V_2\alpha$ is also fixed by V_1, because $V_1V_2\alpha = V_2V_1\alpha = V_2\alpha$. Hence $V_2\alpha = \alpha$, since V_1 has only the fixed point α. Thus V_2 fixes α and we shall prove it fixes nothing else. If not, suppose

† In detail, we select a nonsingular real matrix A so that $A\xi = \infty$, where ξ is the common fixed point, and consider $V_i' z = AV_iA^{-1}z = z + \lambda_i$, $i = 1, 2$. Since V_1' and V_2' commute, so do V_1 and V_2.

$V_2\beta = \beta$, $\beta \neq \alpha$; then V_2 also fixes $V_1\beta$. Since V_2 has at most two fixed points, we must have $V_1\beta = \beta$ or $V_1\beta = \alpha$. The latter is impossible since the one-to-one mapping V_1 already sends α into α. We conclude that β is a second fixed point of V_1, contrary to the assumption that V_1 is parabolic. The theorem is therefore proved in this case.

Next, let V_1 have the distinct fixed points α, β. Then $V_2\alpha$, $V_2\beta$ are also fixed points of V_1, so either V_2 fixes both α and β or else interchanges them. In the first case the theorem is proved. Let us therefore assume $V_2\alpha = \beta$, $V_2\beta = \alpha$, and prove this case is impossible.

It is clear that V_2^2 has the fixed points α and β. Now V_2 (and hence V_2^2) has a fixed point γ, and γ is different from α and β since the latter are not fixed points. It follows that V_2^2 has three distinct fixed points and so is the identity. We deduce that V_2 is elliptic and therefore has a fixed point in H. Let γ be this fixed point; the other fixed point is $\bar{\gamma}$.

We now have V_1 with fixed points α and β, and V_2 with fixed points γ, $\bar{\gamma}$, of which γ is in H. Because of the commutativity, $V_1\gamma$ is a fixed point of V_2. Since $V_1\gamma$ is either γ or $\bar{\gamma}$ and $V_1\gamma$ is in H, we see that γ is fixed by V_1. Then γ must be either α or β, but we saw above that γ is different from both α and β. This contradiction completes the proof.

An immediate corollary is that the theorem is true for two linear transformations that preserve U.

Exercise 1. If U, V are nonparabolic with common fixed point α and the second fixed points distinct, $UVU^{-1}V^{-1}$ is parabolic with fixed point α and is not the identity.

2. Real Discontinuous Groups

We shall study subgroups of Ω_R having a certain property known as discontinuity. Such subgroups will be called *real discontinuous groups.*

Almost all properties of real discontinuous groups are invariant under a homeomorphism of H on U, and hence hold for discontinuous subgroups of Ω_Q. As a matter of convenience we shall sometimes carry out the proofs of such properties in U instead of in H. Discontinuous subgroups of Ω_R or of Ω_Q are known as *Fuchsian groups.*

From now on all groups will be real groups (subgroups of Ω_R) unless the contrary is explicitly stated. It should be kept in mind that the elements of a real group act on all of Z.

2A. We say $\alpha \in$ Z is a *limit point* of the group Γ if there is a $z \in$ Z and a sequence of *distinct* $V_n \in \Gamma$ such that $V_n z \to \alpha$. We denote by L the set of all limit points of Γ and call it the *limit set* of Γ. If z is not a limit point, we say it is an *ordinary point* of Γ, and we denote the set of all ordinary points by $\mathcal{O}$. Thus L and $\mathcal{O}$ are complements of each other in Z.

It is important to note that the images $\{V_n z\}$ need not be distinct. A point that is a fixed point of infinitely many different $V \in \Gamma$ is a limit point of Γ.

DEFINITION. Γ is a discontinuous group if and only if $\mathcal{O}$ is not empty. Γ is discontinuous in a set S if $S \subset \mathcal{O}$.

We say z_1 and z_2 are *equivalent* (or Γ-equivalent) if and only if there is a $V \in \Gamma$ such that $Vz_1 = z_2$. This is an equivalence relation, which therefore partitions Z into disjoint equivalence classes or *orbits*. The orbit of z is the set consisting of z and all points that are equivalent to z by elements of Γ; we denote it by Γz. Thus

$$\Gamma z = \{Vz \mid V \in \Gamma\}.$$

An orbit consists entirely of real points or entirely of nonreal points.

If Γ is discontinuous, no orbit can accumulate at an ordinary point.

2B. THEOREM. Every $V \in \Gamma$ maps L onto L and maps $\mathcal{O}$ onto $\mathcal{O}$.

The first statement implies the second. For if $V(\mathcal{O})$ intersects L, $\mathcal{O}$ intersects $V^{-1}(\mathrm{L})$, a contradiction.

To prove the first statement, let $\lambda \in \mathrm{L}$ and let V_n be distinct elements of Γ such that $V_n z \to \lambda$. Then $VV_n z \to V\lambda$. The $\{VV_n\}$ are distinct since the $\{V_n\}$ are. Hence $\lambda \in \mathrm{L}$ implies $V\lambda \in \mathrm{L}$, or $V(\mathrm{L}) \subset \mathrm{L}$. Since this is true for each $V \in \Gamma$, in particular for V^{-1}, we have $V^{-1}\mathrm{L} \subset \mathrm{L}$, or $\mathrm{L} \subset V(\mathrm{L})$.

2C. THEOREM. If Δ is a subgroup of Γ then $\mathcal{O}(\Gamma) \subset \mathcal{O}(\Delta)$. In particular, a subgroup of a discontinuous group is discontinuous. If $(\Gamma : \Delta)$ is finite, $\mathcal{O}(\Gamma) = \mathcal{O}(\Delta)$. Hence any group containing a discontinuous group as a subgroup of finite index is itself discontinuous.

The first assertion is equivalent to $\mathrm{L}(\Delta) \subset \mathrm{L}(\Gamma)$, which is obvious. Suppose $\Gamma = \Delta A_1 + \Delta A_2 + \cdots + \Delta A_s$. If $\lambda \in \mathrm{L}(\Gamma)$, we have $V_n z \to \lambda$, $V_n \in \Gamma$. Write $V_n = D_n A_{i_n}$, where $D_n \in \Delta$ and $1 \leqslant i_n \leqslant s$. Let i be an i_n occurring infinitely often in $\{A_{i_n}\}$; then $D_m A_i z \to \lambda$ on an infinite subsequence (m) and the D_m are distinct. Thus $\lambda \in \mathrm{L}(\Delta)$; that is, $\mathrm{L}(\Gamma) \subset \mathrm{L}(\Delta)$. From the first statement of the theorem we have $\mathrm{L}(\Delta) \subset \mathrm{L}(\Gamma)$, whence $\mathrm{L}(\Gamma) = \mathrm{L}(\Delta)$.

Exercise 1. For any linear transformation A, belonging to Γ or not, $\mathcal{O}(A\Gamma A^{-1}) = A\mathcal{O}(\Gamma)$.

2D. We give some examples.

(1) $\Gamma =$ transitive group. By a transitive group we mean a group Γ that contains only one orbit; that is, $\Gamma z = \mathrm{Z}$. Obviously a transitive group cannot be real nor can it be discontinuous. Now Ω_C is transitive. For

$S_x = (0\ -1 \mid 1\ -x) \in \Omega_C$ and $S_x(x) = \infty$; hence $S_y^{-1}S_x(x) = y$ for any pair (x, y). Thus Ω_C is not discontinuous.

(2) $\Gamma =$ finite group. Here Γ is discontinuous everywhere, for Γ does not contain an infinite sequence of distinct elements.

(3) $\Gamma =$ cyclic group $= \{T\}$. (In general, by $\{A, B, C, ...\}$ we mean the group generated by $A, B, C, \cdots$.) Here there are several possibilities. First, let T be elliptic; we can then bring it to the form† $z' = ze(\theta)$. If $\theta =$ rational, $\theta = p/q$, then $T^q = I$ and the group is finite, hence discontinuous everywhere. Suppose θ irrational; then the numbers $e(m\theta)$ for integral m are all different, for $e(m\theta) = e(n\theta)$ with $m \neq n$ implies that $(m - n)\theta$ is an integer. Thus all powers of T are distinct. The infinite set $\{e(m\theta), m =$ integer$\}$ has an accumulation point on the unit circle, and $e(m_j\theta) \to e(\varphi)$, say, on a certain sequence $m_j \to \infty$. Hence $T^{m_j}(ze(-\varphi)) \to z$ for each complex z and therefore Γ is not discontinuous.

Next let T be nonelliptic. All powers of T are distinct and a fixed point of T is a fixed point of all its powers. Hence each fixed point of T is a limit point of Γ. We can see from 1B. Exercise 2 that for integers $k_n \to \infty$, $T^{k_n}z$ tends to a fixed point of T. Hence L consists of the fixed points of T.

Among the nonelliptic cyclic groups may be mentioned the *simply periodic* group, generated by a translation $z \to z + \lambda$.

$$(4) \qquad \Gamma = \left\{ V = \begin{pmatrix} a & b \\ c & d \end{pmatrix}, \qquad ad - bc = 1,\ a, b, c, d \text{ integers} \right\}.$$

This is the famous modular group. Since V has real coefficients, it maps H on itself (1C). We shall see later (Theorem 2F) that Γ is discontinuous in both the upper and lower half-planes. On the other hand every real number $x \in$ L. Indeed, let b_n, d_n be relatively prime integers such that $b_n/d_n \to x$ with $\{b_n/d_n\}$ distinct. We then solve the equation $a_n d_n - b_n c_n = 1$ in integers a_n, c_n. Thus $V_n = (a_n\ b_n \mid c_n\ d_n)$ are distinct elements of Γ and $V_n(0) \to x$.

For the modular group, then, L $=$ E.

(5) $\Gamma =$ subgroup of the modular group. Of the great variety of subgroups we mention at this point only one, namely, $\Gamma(n)$, the set of all modular transformations

$$\begin{pmatrix} a & b \\ c & d \end{pmatrix} \equiv \pm \begin{pmatrix} 1 & 0 \\ 0 & 1 \end{pmatrix} \pmod{n}.$$

$\Gamma(n)$ is called the *principal congruence subgroup of level n*; it is clearly a normal subgroup of finite index. From this point of view the modular group may be denoted by $\Gamma(1)$.

Exercise 1. A discontinuous group contains no elliptic elements of infinite order.

† We write $e(\theta)$ for $e^{2\pi i\theta}$.

2E. LEMMA. Let $V_n, V \in \Omega_C$. If $V_n \to V$ in the sense of elementwise convergence, $V_n z \to Vz$ for each $z \in Z$.

If $V_n \to V$, $V^{-1}V_n = W_n \to I$. Let $W_n = (a_n\, b_n \mid c_n\, d_n)$ with $a_n \to 1$, $d_n \to 1$, $b_n \to 0$, $c_n \to 0$. The function $W_n z$ is a rational function of its coefficients and is therefore continuous in a_n, b_n, c_n, d_n for finite z, since $c_n z + d_n \to 1$ as $n \to \infty$. Also $W_n \infty = a_n/c_n \to \infty$. Hence $W_n z \to z$, or $V_n z \to Vz$.

THEOREM. A discontinuous group is countable.

With each element $(a\, b \mid c\, d) \in \Gamma$ we associate in a one-to-one manner a point with coordinates (a, b, c, d) in a euclidean 4-space. The group Γ then corresponds to a certain set S in this space. If Γ is uncountable, so is S, and S contains one of its limit points (a_0, b_0, c_0, d_0). Hence there is a sequence of distinct V_n and an element V belonging to Γ such that $V_n \to V$. It follows that $V_n z \to Vz$ for every z. Now V is nonsingular, so the equation $Vz = \zeta$ can be solved in z for arbitrary ζ. Thus every $\zeta \in Z$ is a limit point, which contradicts the discontinuity of Γ.

2F. A group is called *discrete* if it contains no convergent sequence of distinct matrices.† By convergence we mean elementwise convergence. An obvious example of a discrete group is the modular group (2D).

LEMMA. A group Γ is discrete if and only if there is no sequence $V_n \to I$, where $\{V_n\}$ is distinct.

If $V_n \to I$, Γ contains a convergent sequence and so is not discrete by definition. If Γ is not discrete, there is a sequence $V_n \to V$, $\{V_n\}$ distinct, where V is a real unimodular matrix, since it is the limit of such matrices. Hence V^{-1} exists and $V_n^{-1} \to V^{-1}$. It follows that $V_{n+1}V_n^{-1} \to VV^{-1} = I$. If $\{V_{n+1}V_n^{-1}\}$ contained only finitely many distinct elements, we would have $V_{n+1}V_n^{-1} = I$ for $n > N$, but this contradicts the fact that $\{V_n\}$ is distinct. Hence on a subsequence (m) we have $W_m = V_{m+1}V_m^{-1} \to I$ with distinct W_m.

THEOREM. A group Γ is discontinuous if and only if it is discrete. For a discrete group $L(\Gamma)$ is real.

That a discontinuous group is discrete is trivial. For if Γ is not discrete, $V_n \to I$ for distinct $\{V_n\}$. Hence $V_n z \to z$ for each z and Γ is nowhere discontinuous.

† Note that it is not required that the sequence converge to a *group element*.

Let us now assume, for the converse, that Γ is discrete but $z_0 = x_0 + iy_0$, $y_0 > 0$, is a limit point of Γ. Then $V_n z \to z_0$ for some z. Here $z \in \mathrm{H}$, for Γ preserves the lower half-plane as well as the real axis. Now Ω_R is transitive (see Exercise 1): there is a real unimodular matrix $A \in \Omega_R$ with $z = Ai$. Then $V_n Ai \to z_0$, or $V_n' i \to z_1$, with $V_n' = A^{-1} V_n A$ and $z_1 = A^{-1} z_0$. Observe that $V_n' \in \Gamma' = A^{-1}\Gamma A$, and that $\mathrm{L}(\Gamma') = A^{-1}\mathrm{L}(\Gamma)$ and $\mathrm{L}(\Gamma)$ are simultaneously real. It is therefore sufficient to show $\mathrm{L}(\Gamma')$ is real.

Let $V_n' = (\alpha_n \beta_n \mid \gamma_n \delta_n)$. Since $z_1 \in \mathrm{H}$ we have

$$\operatorname{Im} V_n' i = \frac{1}{\gamma_n^2 + \delta_n^2} \to \operatorname{Im} z_1 > 0, \qquad |V_n' i|^2 = \frac{\alpha_n^2 + \beta_n^2}{\gamma_n^2 + \delta_n^2} \to |z_1|^2 > 0.$$

From the first relation we deduce that the sets $\{\gamma_n\}$ and $\{\delta_n\}$ are bounded, from the second that $\{\alpha_n\}$ and $\{\beta_n\}$ are bounded. Hence we can select a convergent subsequence $\{V_m'\}$, whence Γ' is not discrete. Therefore Γ is not discrete. The argument would go just as well for a z_0 in the lower half-plane, and this concludes the proof that $\mathrm{L}(\Gamma)$ is real.

COROLLARY. For a real group either $\mathrm{L} = \mathrm{Z}$ or $\mathrm{L} \subset \mathrm{E}$.

The groups of interest to us are therefore the discrete subgroups of Ω_R.

By mapping H on U we see that the theorem holds for subgroups of Ω_Q: a subgroup of Ω_Q is discontinuous if and only if it is discrete, and in that case $\mathrm{L} \subset \mathrm{Q}$.

Exercise 1. Ω_R is transitive.

Exercise 2. Let L_1, L_2 be lines or circles orthogonal to E. There is an element of Ω_R mapping L_1 on L_2.

2G. We define the *group of stability* or *stabilizer* of a point z to be $\{T \in \Gamma \mid Tz = z\}$ and denote it by Γ_z. We can easily check that $\Gamma_{Az} = A\Gamma_z A^{-1}$ for $A \in \Gamma$.

As an aid in the study of the stabilizer we note the following

THEOREM. An abelian subgroup of a discrete group Γ belonging to $\bar{\Omega}_R$ is cyclic.†

Let Δ be an abelian subgroup of Γ. Since all elements of Δ commute, they all have the same fixed point set (1E), hence all are parabolic, or all elliptic, or all hyperbolic. (Δ cannot contain both hyperbolic and elliptic elements, for the former have real fixed points while the latter do not.) Suppose Δ consists entirely of parabolic elements (and the identity), which we may assume (by

† The theorem is not necessarily true for subgroups of Ω_R. Thus the group generated by $-I$ and $(1\ 1 \mid 0\ 1)$ is abelian and discrete but not cyclic.

transforming the group) to be translations $Tz = z + \lambda$. If we denote the set of periods λ by Λ, then Λ is a discrete module; that is, $\lambda_1 - \lambda_2 \in \Lambda$ whenever $\lambda_1, \lambda_2 \in \Lambda$, and Λ contains no convergent sequence of its elements. The latter property comes about because $\lambda_n \to \lambda$ would imply $T_{\lambda_n}z \to Tz$, which is impossible in a discrete group.

Since 0 is not an accumulation point of Λ, and since Λ contains the negative of each of its elements, Λ contains a smallest positive element d. For $\lambda \in \Lambda$ we have $\lambda = qd + r$, where q is an integer and $0 \leq r < d$. But $r = \lambda - qd \in \Lambda$, and $r > 0$ would imply that d is not the smallest positive element. Hence $\lambda = qd$; that is, Λ is the set of integral multiples of d. It follows that Δ is generated by $T_d z$.

If Δ consists of hyperbolic elements, we write (after transforming the group so that the fixed points are at 0 and ∞), $\log Tz = \log \kappa + \log z$ and apply the same reasoning to $\{\log \kappa\}$. If Δ consists of elliptic elements, use $i \log \kappa$.

2H. THEOREM†. If Γ is discrete, Γ_z is finite cyclic or the identity when $z \in$ H, infinite cyclic or the identity when $z \in$ E.

In any case Γ_z is discrete. Let $z \in$ H; an element of Γ fixing z must be elliptic unless it is the identity. Hence Γ_z consists entirely of elliptic elements with the same fixed points (z and $\bar{z}$) and is therefore abelian, hence cyclic. It is generated by an element of finite order and so is finite cyclic.

In order to treat the case $z \in$ E we shall need an additional result.

2I. THEOREM. Let $H \in \Gamma$ be hyperbolic and let $V \in \Gamma$ have one and only one fixed point in common with H. Then Γ is not discrete.

We assume without loss of generality that H has fixed points 0 and ∞, the latter being the common fixed point. Then

$$H = \begin{pmatrix} \alpha & 0 \\ 0 & \alpha^{-1} \end{pmatrix}, \qquad |\alpha| \neq 0, 1; \qquad V = \begin{pmatrix} a & b \\ 0 & a^{-1} \end{pmatrix}, \qquad ab \neq 0.$$

We have

$$VH^nV^{-1}H^{-n} = C_n = \begin{pmatrix} 1 & ab(1 - \alpha^{2n}) \\ 0 & 1 \end{pmatrix}$$

and $C_n \in \Gamma$. Hence $C_n \to C = (1\ ab \mid 0\ 1)$ for $n \to \infty (|\alpha| < 1)$ or $n \to -\infty (|\alpha| > 1)$. Since $ab \neq 0$ and $|\alpha| \neq 0, 1$, the C_n are distinct. Hence Γ is not discrete.

COROLLARY. A fixed point of a hyperbolic element of a discrete group Γ is not a fixed point of a parabolic element of Γ. The fixed-point sets of two hyperbolic elements are identical or disjoint.

† See footnote on page 14.

2J. We can now complete the proof of Theorem 2H. Let $z \in \mathrm{E}$ be fixed by a parabolic element. Then z is not fixed by any hyperbolic element and certainly not by an elliptic element. That is, Γ_z consists entirely of parabolic elements with a common fixed point and by the previous reasoning is cyclic. Since a parabolic element is always of infinite order, Γ_z is infinite cyclic. The same reasoning applies to a point fixed by a hyperbolic element. The stabilizer of such a point can contain only hyperbolic elements with a common pair of fixed points.

Exercise 1. If $\Gamma\alpha = \alpha$ for some α, Γ has at most two limit points. [$\Gamma = \Gamma_\alpha$ is cyclic.]

Exercise 2. If $\mathrm{L}(\Gamma) = \{p\}$, then $\Gamma = \Gamma_p$ and is a parabolic cyclic group. [If $\Gamma_p = \emptyset$, then from $V_n z \to p$ we deduce $VV_n \to Vp \neq p$, implying that $\mathrm{L} \neq \{p\}$. Let $\Gamma_p = \{P\}$. Suppose $T \in \Gamma - \Gamma_p$; then TP^nT^{-1} has fixed point $Tp \neq p$ and $Tp \in \mathrm{L}$. Finally, P is parabolic.]

Exercise 3. Let H be a normal subgroup of a discrete horocyclic group Γ. Then $\mathrm{L}(H) \neq \emptyset$. ($\Gamma$ is called horocyclic if $\mathrm{L}(\Gamma) = \mathrm{E}$.)
[In contradiction assume H is a finite group and so contains only elliptic elements. If $D \subset \Gamma$ consists of the $V \in \Gamma$ that commute with a given E in H, then D is finite. Let $V_i \in \Gamma - D$, $i = 1, 2, \cdots$; the transformations $\{V_i E V_i^{-1}\}$ are in H and include infinitely many different ones, a contradiction.]

Exercise 4. Let H be a normal subgroup of a discrete horocyclic group Γ. Then $\mathrm{L}(H) \neq \{p\}$.
[Otherwise $H = \{P\}$, where we may assume $P = (1\ \lambda \mid 0\ 1)$. Then $VPV^{-1} \neq P^m$, where $V = (a\ b \mid c\ d) \in \Gamma$ has $c \neq 0$. If every $V \in \Gamma$ has $c = 0$, show that Γ is not horocyclic.]

Exercise 5. For a subset H of the group G we call

$$N_G(H) = \{x \in G \mid xHx^{-1} = H\}$$

the normalizer of H in G. Calculate the normalizer of a single element of a discrete group Γ; also of a cyclic subgroup of Γ.

2K. The following beautiful result is due to S. Lauritzen. Subsequent proofs were given by J. Nielsen and C. L. Siegel. We reproduce Siegel's proof.

THEOREM. Let Γ be a nonabelian real group consisting of the identity and hyperbolic elements. Then Γ is discrete.

Transform the group so that it contains an element A with fixed points 0 and ∞:

$$A = \begin{pmatrix} \lambda & 0 \\ 0 & \lambda^{-1} \end{pmatrix}, \qquad \lambda > 0, \qquad \lambda \neq 1.$$

We separate the main part of the proof in a

LEMMA. If Γ contains a sequence $V_n = (a_n\, b_n \mid c_n\, d_n) \to I$, then for $n > N$ we have

$$V_n = \begin{pmatrix} \rho_n & 0 \\ 0 & \rho_n^{-1} \end{pmatrix}, \qquad \rho_n^2 \neq 1.$$

Since $V_n \to I$, we have $a_n d_n \to 1$ and $b_n c_n \to 0$. Thus $a_n d_n$ is eventually positive. We now compute the trace of the commutators

$$C_n = AV_nA^{-1}V_n^{-1} = \begin{pmatrix} 1 - b_nc_n(\lambda^2 - 1) & a_nb_n(\lambda^2 - 1) \\ c_nd_n(\lambda^{-2} - 1) & 1 - b_nc_n(\lambda^{-2} - 1) \end{pmatrix},$$

$$D_n = AC_nA^{-1}C_n^{-1} = \begin{pmatrix} 1 + a_nb_nc_nd_n(\lambda^2 - 1)(\lambda^{-2} - 1)(1 - \lambda^2) & \\ & 1 + a_nb_nc_nd_n(\lambda^2 - 1)(\lambda^{-2} - 1)(1 - \lambda^{-2}) \end{pmatrix},$$

as

$$\chi(C_n) = 2 - b_nc_n(\lambda - \lambda^{-1})^2, \; \chi(D_n) = 2 + a_nb_nc_nd_n(\lambda - \lambda^{-1})^4.$$

Now $|\chi(C_n)| \geq 2$; since $b_nc_n(\lambda - \lambda^{-1})^2 \to 0$, we have eventually $\chi(C_n) \geq 2$ and so

$$b_nc_n \leq 0, \qquad n > N.$$

But also $|\chi(D_n)| \geq 2$ and $a_nb_nc_nd_n(\lambda - \lambda^{-1})^4 \to 0$. Hence eventually $\chi(D_n) \geq 2$ and $a_nb_nc_nd_n$ is nonnegative. The positivity of a_nd_n gives

$$b_nc_n \geq 0, \qquad n > N,$$

which, with the above, yields

$$b_nc_n = 0, \qquad n > N.$$

Thus we have $\chi(C_n) = 2$, $n > N$. Therefore $C_n = \pm I$, for Γ contains no parabolic elements. This implies that V_n commutes with A and so has fixed points 0 and ∞ (1E). Hence $b_n = c_n = 0$ for $n > N$ and the lemma is proved.

The proof of the theorem is now easy. If the theorem is false, there is a convergent sequence in Γ and therefore a sequence $V_n \to I$. By virtue of the lemma we may assume $V_n = \begin{pmatrix} \rho_n & 0 \\ 0 & \rho_n^{-1} \end{pmatrix}$ for all n, where $\rho_n^2 \neq 1$.

Since Γ is not abelian, it contains an element B that does not have 0 and ∞ as fixed points. B is then of the form

$$B = \begin{pmatrix} \alpha & \beta \\ \gamma & \delta \end{pmatrix}, \qquad |\alpha + \delta| > 2, \qquad \beta^2 + \gamma^2 > 0$$

for B is hyperbolic. Consider

$$X_n = V_nBV_n^{-1}B^{-1} = \begin{pmatrix} \cdot & (\rho_n^2 - 1)\alpha\beta \\ (\rho_n^{-2} - 1)\gamma\delta & \cdot \end{pmatrix}.$$

Obviously $X_n \to I$; hence by the lemma

$$X_n = \begin{pmatrix} \cdot & 0 \\ 0 & \cdot \end{pmatrix}. \qquad n > N.$$

Since $\rho_n^2 \neq 1$, $\rho_n^{-2} \neq 1$, comparison gives $\alpha\beta = \gamma\delta = 0$.

Suppose $\alpha = 0$. Since then $\gamma \neq 0$—because $\alpha\delta - \beta\gamma = 1$—this gives $\delta = 0$, hence $\alpha + \delta = 0$, a contradiction. Therefore $\alpha \neq 0$, so that $\beta = 0$. Then $\delta \neq 0$, implying $\gamma = 0$. Hence $\beta^2 + \gamma^2 = 0$, again a contradiction. There is no sequence $V_n \to I$ and Γ is discrete.

3. The Limit Set of a Discrete Group

3A. By definition a limit point of a group Γ is the limit of a *sequence* of images $V_n z$, $V_n \in \Gamma$. The following theorem considers the *orbits* Γz in relation to the limit set L. On the way we shall discover that L is a closed set.

THEOREM. To each $\lambda \in$ L there is a $\lambda' \in$ L such that the orbit Γz is dense at λ provided $z \neq \lambda$, $z \neq \lambda'$.

Let us first dispose of the case in which λ is a fixed point, necessarily of a parabolic or hyperbolic element of Γ. If $\lambda = T\lambda$, T parabolic, and $z \neq \lambda$, then $T^n z \to \lambda$, $n \to \infty$, and $\{T^n z\}$ is a set of distinct points since λ is the only fixed point of T. Hence Γz is dense at λ for $z \neq \lambda$; that is, in this case $\lambda = \lambda'$. Let $\lambda = T\lambda$ with T hyperbolic and let λ' be the other fixed point of T. Then $T^n z \to \lambda$ either for $n \to \infty$ or for $n \to -\infty$, provided $z \neq \lambda, \lambda'$. Hence we may assume from now on that λ is not a fixed point.

It is convenient to prove the theorem for groups that preserve the unit disk. We still call the transformed group Γ and its limit set L, which is now a subset of the unit circle. The elements of Γ have the form

$$V = \begin{pmatrix} a & \bar{b} \\ b & \bar{a} \end{pmatrix}, \qquad a\bar{a} - b\bar{b} = 1.$$

Note that $b = 0$ for only finitely many $V \in \Gamma$, for $b = 0$ implies that V fixes ∞, and ∞ is an ordinary point for Γ.

Define

$$A = \left\{x \mid \exists \begin{pmatrix} x & \cdot \\ \cdot & \cdot \end{pmatrix} \in \Gamma\right\}, \qquad B = \left\{y \mid \exists \begin{pmatrix} \cdot & \bar{y} \\ \cdot & \cdot \end{pmatrix} \in \Gamma\right\}.$$

That is, A is the set of entries in the upper left-hand corner of the matrices of Γ.

LEMMA. If Γ is discrete, the sets A and B have no finite points of accumulation.

Suppose in contradiction there are distinct $V_n \in \Gamma$ such that $b_n \to b$ (finite). Then $|b_n|^2 \to |b|^2$ and from the determinant condition we get $|a_n|^2 \to 1 + |b|^2$. Hence the sets $\{a_n\}$, $\{b_n\}$ are bounded and we can extract a convergent subsequence $\{V_p\}$. Thus Γ is not discrete. This proves the property for B and the proof for A is the same.

COROLLARY. There exists a positive number $\tilde{b}$ such that for $V = (a\,\bar{b} \mid b\,\bar{a}) \in \Gamma$ either $b = 0$ or $|b| \geq \tilde{b} > 0$.

3B. In order to get a hold on L we introduce a new set,

$$\mathrm{L}_\infty = \{\alpha \mid \alpha = \lim_{n\to\infty} V_n\infty, \text{ where } V_n \in \Gamma, \text{ distinct}\}.$$

Obviously $\mathrm{L}_\infty \subset \mathrm{L}$.

LEMMA. (1) $V\mathrm{L}_\infty = \mathrm{L}_\infty$ for each $V \in \Gamma$.

(2) L_∞ is closed.

The proof of (1) is the same as that of Theorem 2B. To prove (2) let λ be an accumulation point of L_∞ and λ_k a member of L_∞ for which $|\lambda - \lambda_k| < k^{-1}$, $k = 1, 2, \cdots$. Choose distinct elements V_{n_k} such that $|V_{n_k}\infty - \lambda_k| < k^{-1}$. Then $|\lambda - V_{n_k}\infty| < 2k^{-1} \to 0$ with $k \to \infty$. Hence $\lambda \in \mathrm{L}_\infty$.

THEOREM. $\mathrm{L} = \mathrm{L}_\infty$.

We need show only $\mathrm{L} \subset \mathrm{L}_\infty$. Suppose $\lambda \in \mathrm{L}$, then for some z,

$$\lambda = \lim_{n\to\infty} V_n z, \qquad V_n = \begin{pmatrix} a_n & \bar{b}_n \\ b_n & \bar{a}_n \end{pmatrix}, \qquad \text{distinct.}$$

We may assume all $b_n \neq 0$. Either

(1) $|b_n z + \bar{a}_n| \geqslant 1$ for $n > N$, or

(2) $|b_p z + \bar{a}_p| < 1$ on a subsequence $(p) \to \infty$.

In case (2) we see at once that $z \in \mathrm{L}_\infty$, for $|z + a_p/b_p| = |z - V_p^{-1}\infty| < |1/b_p| \to 0$ by Lemma 3A. From $V_n\mathrm{L}_\infty = \mathrm{L}_\infty$ we deduce $V_n z$ is in L_∞ for each n. It follows that $\lambda = \lim V_n z$ lies in L_∞, since L_∞ is closed.

In case (1) we have

$$|V_n z - V_n\infty| = \left|\left(\frac{a_n}{b_n} - \frac{1}{b_n(b_n z + \bar{a}_n)}\right) - \frac{a_n}{b_n}\right| = \frac{1}{|b_n|}\frac{1}{|b_n z + \bar{a}_n|} \leq \frac{1}{|b_n|},$$

so that $\lambda - V_n\infty = (\lambda - V_n z) + (V_n z - V_n\infty) \to 0$ and $\lambda \in \mathrm{L}_\infty$.

3C. An immediate corollary of these results is the fundamental

THEOREM. The limit set of a discrete group is a closed set. The set of ordinary points is open.

Exercise 1. Complete the details of the following proof of Theorem 2E. Write $\mathcal{O}$ as a denumerable union of *compact* K_n; this is possible since $\mathcal{O}$ is open. Enumerate Γ by counting, for each n, those $V \in \Gamma$ for which $V\alpha$ lies in K_n, with α a fixed point of $\mathcal{O}$. Show that $V\alpha \in K_n$ for only finitely many V.

3D. We return to the proof of Theorem 3A. Let λ be a nonfixed point belonging to L; as we have just seen $\lambda \in L_\infty$ and so

$$V_n\infty \to \lambda, \qquad V_n = \begin{pmatrix} a_n & \bar{b}_n \\ b_n & \bar{a}_n \end{pmatrix}, \qquad \text{distinct.}$$

Hence, as before,

$$| V_n z - V_n \infty | = \frac{1}{| b_n |^2} \frac{1}{| z + \bar{a}_n/b_n |} . \tag{12}$$

The set $\{-\bar{a}_n/b_n\} = \{V_n^{-1}\infty\}$ is bounded, for it cannot accumulate at the ordinary point ∞. Hence it admits an accumulation point λ' and there is a subsequence V_p on which

$$V_p\infty \to \lambda, \qquad V_p^{-1}\infty \to \lambda'.$$

Obviously λ' is a limit point.

If $z \neq \lambda'$, then z is isolated from the set $\{V_p^{-1}\infty\}$. Hence

$$\left| z + \frac{\bar{a}_p}{b_p} \right| \geq m > 0$$

for $p > N$. It follows from (12) and Lemma 3A that

$$| V_p z - V_p \infty | \leq m^{-1} | c_p |^2 \to 0.$$

Thus

$$V_p z \to \lambda.$$

Now if $\{V_p z\}$ contains only finitely many distinct elements, then $V_p z = \lambda$ for $p \geq N$ and λ is a fixed point of $V_{p+1}V_p^{-1}$, contrary to our present assumption. Hence Γz is dense at λ for $z \neq \lambda'$. This concludes the proof.

The argument shows incidentally the following: *if S is a compact set that avoids λ and λ', there exist $V_n \in \Gamma$ such that $V_n z \to \lambda$ uniformly on S.*

Exercise 1. $L = d(\Gamma z)$ for each $z \in \mathcal{O}$, where $d(A)$ is the derived set of A (set of points of accumulation of A).

Exercise 2. If L contains more than two points, $L = d(\Gamma z)$ for each $z \in Z$.

3E. The following theorems are consequences of Theorem 3A.

THEOREM 1. If L contains more than two points, it is a perfect set.

Since L is closed, we have only to show that no point of L is isolated. Let $\lambda \in$ L. By hypothesis there are two other distinct points μ, $\nu \in$ L different from λ. According to Theorem 3A either $\Gamma\mu$ or $\Gamma\nu$ is dense at λ. Since $\Gamma\mu$, $\Gamma\nu \in$ L, λ is an accumulation point of L.

A limit set is therefore empty, one point, two points, or a perfect set.

THEOREM 2. L $=$ E or L is a nowhere dense subset of E.

If L $\neq$ E, let $\alpha \in$ E $-$ L; α is an ordinary point. Let $\lambda \in$ L. Then $\Gamma\alpha$ is dense at λ and $\Gamma\alpha \subset$ E. Every real neighborhood of λ contains points in the complement of L, so L is nowhere dense in E.

The groups for which L $=$ E are called groups of the first kind; those for which L $\neq$ E are called groups of the second kind. Groups of the first kind are also called *horocyclic groups*, in translation of the German name *Grenzkreisgruppen.*

THEOREM 3. If S is a closed set containing at least two points such that $V(S) \subset S$ for all $V \in \Gamma$, then $S \supset$ L.

The assumption that $\lambda \in$ L does not lie in S implies there is a neighborhood N of λ that does not intersect S. Now S contains distinct points z_1, z_2 and by Theorem 3A at least one of them has images lying in N and therefore outside S. This contradicts the hypothesis that S is mapped into itself by each V in Γ.

The theorem may be phrased as follows: when Γ contains more than one limit point, L is the smallest closed Γ-invariant set containing at least two points.

Exercise 1. Let H be a normal subgroup of the discrete horocyclic group Γ. Then H is horocyclic.
[L$(H) \subset$ L(Γ) and we need L$(H) \supset$ L(Γ). Use Theorem 3. L(H) has more than one point by 2J, Exercises 3, 4. To show invariance, choose $\lambda \in$ L(H), $V \in \Gamma$. There exist $h_n \in H$ such that $h_n V^{-1} z \to \lambda$ for a $z \in$ H. Hence $V h_n V^{-1} z \to V\lambda$ and $V h_n V^{-1} \in H$; thus $V\lambda \in$ L(H) and L(H) is Γ-invariant.]

Exercise 2. L is the closure of the set of fixed points of all elements of Γ.

Exercise 3. If Γ contains hyperbolic elements, L is the closure of the set of fixed points of all hyperbolic elements of Γ.

Exercise 4. If Γ contains both hyperbolic and parabolic elements, L is the closure of the set of fixed points of all parabolic elements of Γ.

[Let H be hyperbolic, P parabolic, and $P\xi = \xi$. Then H does not fix ξ and HPH^{-1} is parabolic with fixed point $H\xi \neq \xi$.]

4. The Fundamental Region

4A. Let Γ be a real discrete group. The relation of Γ-equivalence partitions H into disjoint orbits Γz. A subset F of H that contains exactly one point from each orbit is called a *fundamental set* for Γ *relative to* H. Thus F is a fundamental set if and only if:

(1) No two distinct points of F are Γ-equivalent.

(2) Every $z \in$ H is Γ-equivalent to a point of F.

By the axiom of choice, a fundamental set exists for every group of transformations of H, but it is in no way unique. Thus if $A \subset F$ and $V \in \Gamma$, then $(F - A) \cup VA$ is also a fundamental set. It is evident that a fundamental set need have no convenient topological properties.

But the familiar groups all have very simple fundamental sets. The simply periodic group has a fundamental set consisting of a strip with one boundary adjoined; the doubly periodic group, a parallelogram with two adjacent open sides and their common vertex adjoined. Similar constructions can be made for the cyclic groups, the modular group, etcetera. In each case the fundamental set is an open set with some of its boundary points adjoined, and the complete boundary consists of line segments or circular arcs.

We shall show that every discrete group admits a fundamental set of this type.

A fundamental set cannot be open, for it must contain points equivalent to points on its own boundary. Since it is convenient to work with either open or closed sets, we shall modify the concept slightly and make the following

DEFINITION. An open subset R of H is a fundamental region for Γ provided:

(1) No two distinct points of R are Γ-equivalent.

(2) Every point of H is Γ-equivalent to a point of $\bar{R}$.

From a fundamental region we can easily determine a fundamental set (see Exercise 1).

The principal tool in the existence proof to follow will be the construction of a model of the plane of hyperbolic geometry. This model is known as the Poincaré half-plane. Its applications go much further than the construction of fundamental regions.

Exercise 1. If R is a fundamental region for Γ, there is a fundamental set F such that $R \subset F \subset \bar{R}$.

Exercise 2. A subgroup of Ω_R that possesses a fundamental region is discontinuous.

4B. Hyperbolic Geometry. Plane hyperbolic geometry is obtained from plane euclidean geometry by replacing the parallel postulate by the following axiom: *Through a given point not on a given line there passes more than one line that does not meet the given line.*

In Poincaré's model H represents the hyperbolic plane. A point in the hyperbolic plane is represented by a point in H. A line is represented by that arc of a circle orthogonal to E which lies in H; we include explicitly straight lines orthogonal to E. We call these elements the *H*-plane, *H*-point, and *H*-line, respectively. We define *H*-angle measure to be the same as euclidean-angle measure.

It is seen that the axioms of hyperbolic geometry are fulfilled in the model. For example, two *H*-lines intersect, if at all, in a single point, and through any two distinct points of H there passes exactly one *H*-line. These are simply restatements of familiar facts from euclidean geometry. As for the modified parallel axiom, consult Figure 1. Through the point P there pass two lines,

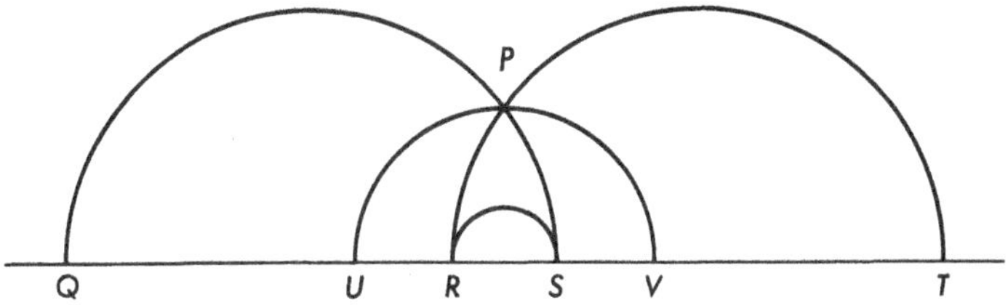

FIG. 1.

QPS and RPT, which do not meet the line RS. Obviously any line within the angle QPR, such as UV, also fails to meet RS.

Since all axioms of euclidean geometry except the parallel axiom are valid in hyperbolic geometry, all theorems not depending on the parallel axiom are valid in hyperbolic geometry. Thus the *H*-line that cuts the *H*-line ab orthogonally at its mid-point is the locus of points equidistant from a and b. This fact is basic to our later considerations and we shall give a second proof presently. Among the notable theorems that change is the one stating the angle sum of a triangle. In hyperbolic geometry this sum is always less than π and it can be zero.

In hyperbolic geometry, as in euclidean geometry, a distance is defined, and distance is invariant under a group of "rigid motions." The rigid motions also preserve straight lines. In our model the group of rigid motions will be the group Ω_R. Distance will be defined by its differential

$$ds = \frac{|d\tau|}{y}, \qquad \tau = x + iy. \tag{13}$$

To show invariance of ds under Ω_R, let $T = (a\ b \mid c\ d) \in \Omega_R$. Then from

$$y' = \frac{y}{|c\tau + d|^2}, \qquad d\tau' = \frac{d\tau}{(c\tau + d)^2}, \qquad \tau' = T\tau = x' + iy'$$

we get

$$\frac{|d\tau'|}{y'} = \frac{|d\tau|}{y}, \tag{14}$$

as asserted. The distance between two points is then defined as $\int ds$, the integral being extended along the H-line joining the points. By this definition we obviously have

$$d(a, b) = d(Ta, Tb), \qquad T \in \Omega_R, \qquad a, b \in \mathrm{H}$$

where d denotes H-distance. Indeed, if C is the H-segment connecting a and b, then T maps C on C', the H-segment connecting Ta and Tb, and (14) yields

$$\int_a^b \frac{|d\tau|}{y} = \int_{Ta}^{Tb} \frac{|d\tau'|}{y'}, \qquad \tau' = T\tau.$$

That is, H-distance is invariant under mappings of Ω_R. We also have $d(A, B) + d(B, C) = d(A, C)$ if A, B, C are colinear points. That is, distance is additive along H-lines.

We already know that H-lines are mapped into circles or straight lines by elements of Ω_R. Every such element is a conformal transformation and maps E into itself; it follows that the image of an H-line is a circle or straight line orthogonal to E and is therefore an H-line.

Let us now prove that H-lines are geodesics; this will show in particular that *H-distance satisfies the triangle inequality* and so has the properties of a metric. Because Ω_R is transitive with respect to H-lines (see 2F, Exercise 2), we have to do this only for a line segment lying on the imaginary axis, which we take to be

$$L\colon \tau = iy, \qquad 1 \leq y \leq y_0.$$

Let L' be a differentiable curve lying in H and connecting the endpoints of L, namely,

$$L' = \tau = x + iy, \qquad x = x(t), \qquad y = y(t), \qquad 0 \leq t \leq 1,$$

$$x(0) = x(1) = 0, \qquad y(0) = 1, \qquad y(1) = y_0.$$

Denote the H-length of L' by $|L'|$. Then

$$|L'| = \int_{L'} \frac{|d\tau|}{y} = \int_0^1 \frac{\sqrt{(x'(t))^2 + (y'(t))^2}}{y(t)}\, dt \geq \int_0^1 \frac{|y'(t)|}{y(t)}\, dt$$

$$\geq \int_0^1 \frac{y'(t)}{y(t)}\, dt = \log \frac{y(1)}{y(0)} = \log y_0 = \int_1^{y_0} \frac{dy}{y} = |L|.$$

The differential of area is defined by

$$d\sigma = \frac{dx\,dy}{y^2}. \tag{15}$$

To demonstrate its invariance, recall that

$$dx'\,dy' = \frac{\partial(x', y')}{\partial(x, y)}\,dx\,dy,$$

where $x' + iy' = \tau' = T\tau$. Since τ' is an analytic function of τ we have, from the Cauchy-Riemann equations,

$$\frac{\partial(x', y')}{\partial(x, y)} = |\,T'(\tau)|^2 = \frac{1}{|\,c\tau + d\,|^4} = \frac{y'^2}{y^2},$$

and the result follows.

It is evident that the distance $d(z_0, z) \to \infty$ as z approaches the real axis. Of two equal euclidean line segments, therefore, the one nearer the real axis has greater hyperbolic length.

We have seen that H is a metric space under H-distance. The topology defined by this metric is easily seen to be equivalent to the usual euclidean topology. That is, every open set in one topology contains an open set in the other topology. A basis for the open sets in the H-topology consists of the H-disks

$$S(\tau_0, r) = \{w \in \mathrm{H} \mid d(w, \tau_0) < r, \qquad r > 0\}.$$

Moreover, an H-disk is also a euclidean disk, as we now show.

Let $\tau_0 \in \mathrm{H}$ and let M be an elliptic element of Ω_R with fixed points τ_0 and $\bar{\tau}_0$ and multiplier κ, $|\,\kappa\,| = 1$. As we saw in 1C, the fixed circles of M are euclidean circles orthogonal to the circles passing through τ_0 and $\bar{\tau}_0$. Let K be a fixed circle and ζ, ζ' two points on K with $\zeta' = M\zeta$. Since κ is arbitrary, any pair of points on K satisfy such an equation. Then $d(\tau_0, \zeta) = d(M\tau_0, M\zeta) = d(\tau_0, \zeta')$. That is, all points on K are at the same H-distance from τ_0, say r. On the other hand if $d(\tau_0, \omega) = r$, draw the euclidean circle through τ_0, ω, and $\bar{\tau}_0$; it will be orthogonal to K and will cut K in a point ω', and it is clear that $\omega' = \omega$. This shows that a "hyperbolic circle"—the locus of points at a fixed H-distance from a fixed center—is a euclidean circle, and justifies our statement about H-disks.

These remarks enable us to interpret the elements of Ω_R as noneuclidean rigid motions. Suppose T is elliptic with fixed point τ_0 and multiplier $\kappa = \exp i\theta$. As we have just seen, a fixed circle K of T is an H circle about τ_0, and T moves an H-ray through τ_0 perpendicular to K into another H-ray making an angle θ with the first. We may then call T a (hyperbolic) rotation about τ_0. If T is hyperbolic, it maps the H-line connecting the fixed points into itself, and in consequence is called a (hyperbolic) translation. The limiting case of an

H-rotation, in which the complex conjugate fixed points approach the real axis, is a parabolic element of Ω_R, and is called a limit rotation.

Let us now give another proof that the hyperbolic perpendicular bisector of the H-line ab is the locus of all points in H that are H-equidistant from a and b.

The result is invariant under mappings of Ω_R, which preserve distance and angle. By the transitivity of Ω_R (2F, Exercise 1) we may therefore assume the points $z_1 = x + iy$, $z_2 = -x + iy$ to lie on $Q'(x^2 + y^2 = 1, y > 0)$. In fact Q' is the H-line z_1z_2 and its perpendicular bisector is the imaginary axis, for it is clear by symmetry that

$$\int_{z_1}^{i} \frac{|d\tau|}{y} = \int_{i}^{z_2} \frac{|d\tau|}{y}.$$

Hence we must show the imaginary axis is the locus of points H-equidistant from z_1 and z_2.

Let C_1 be an H-circle of radius r about z_1 that intersects the imaginary axis. Under the mapping $z \to -\bar{z}$, $x \to -x$ while y remains invariant; hence H-length is preserved. It follows that under this transformation C_1 goes into the H-circle C_2 about z_2 of radius r, and C_1, C_2 intersect on the imaginary axis in two points that are both equidistant from z_1 and z_2. Conversely, a point z equidistant from z_1 and z_2 determines two H-circles that, by the preceding reasoning, intersect on the imaginary axis. This establishes the result.

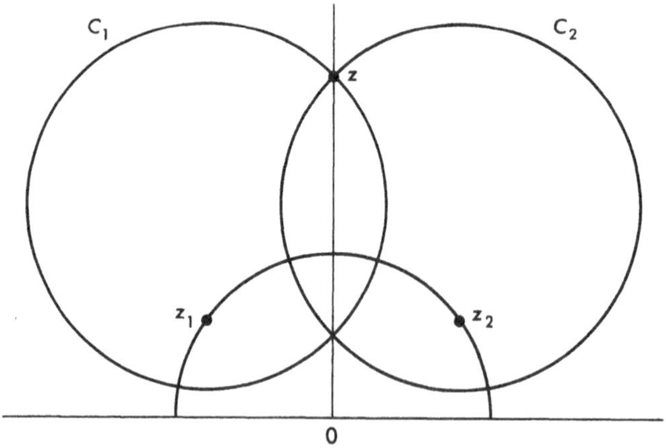

FIG. 2.

The notion of convexity carries over to H-geometry. An *H-convex set* is a subset of H that contains the H-line joining any two of its points. A region that is an H-convex set is called an *H-convex region.* The following properties are immediate: an H-convex set is connected; the intersection of H-convex sets (regions) is an H-convex set (region).

Exercise 1. The closure of an H-convex region is H-convex. [Show that an H-ray from a fixed interior point meets the boundary of the region in exactly one point.]

As a first application of the hyperbolic geometry let us prove the following fundamental

THEOREM. Let Γ be a discrete group. Each $w_0 \in \mathrm{H}$ that is not fixed by any element of Γ lies in an open set D containing no Γ-equivalent points.

Indeed, define

$$D = \{w \in \mathrm{H} \mid d(w, w_0) < \tfrac{1}{2}\delta\},$$

where $\delta = \inf \{d(w_0, Vw_0) | \; V \in \Gamma, V \neq I\}$. Now $\delta > 0$, otherwise there would be a sequence $V_n w_0 \to w_0$ with different V_n, and w_0 could not be an ordinary point. Suppose now ζ and $\zeta' = V\zeta$, $V \in \Gamma$, both lie in D. We have $d(w_0, Vw_0) \leq d(w_0, \zeta') + d(\zeta', Vw_0)$, or, by the invariance of distance,

$$d(\zeta, w_0) = d(\zeta', Vw_0) \geq d(w_0, Vw_0) - d(w_0, \zeta').$$

Since w_0 is not fixed, we have $d(w_0, Vw_0) \geq \delta$, while $d(w_0, \zeta') < \delta/2$. Hence

$$d(\zeta, w_0) > \delta - \frac{\delta}{2} = \frac{\delta}{2},$$

a contradiction to $\zeta \in D$. Now D is obviously open, and this concludes the proof.

4C. We shall now construct a fundamental region for the real discrete group Γ. Let $w_0 \in \mathrm{H}$ be a point not fixed by any element of Γ. There certainly is such a point, for the elements of Γ are countable in number (2E) and each has at most one fixed point in H. The point w_0, once selected, will not be changed in the course of the following construction. Let $\{V_i, i = 0, 1, 2, \cdots; V_0 = I\}$ be an enumeration of the elements of Γ. The images $w_i = V_i w_0$ are all distinct, for $w_i = w_j$ implies that $V_j^{-1} V_i$ fixes w_0. Thus for $i > 0$, $w_0 w_i$ is an H-segment, not a point.

Denote by λ_i the H-perpendicular bisector of the segment $w_0 w_i$. The line λ_i divides H into two "half-planes"; the one that contains w_0 we call L_i, the complementary one, L_i'. These half-planes are open and do not include the points of λ_i.

LEMMA 1.
$$L_i = \{z \in \mathrm{H} \mid d(z, w_0) < d(z, w_i)\}$$
$$\lambda_i = \{z \in \mathrm{H} \mid d(z, w_0) = d(z, w_i)\}$$
$$L_i' = \{z \in \mathrm{H} \mid d(z, w_0) > d(z, w_i)\}$$

The second statement has been proved in 4B. Since λ_i, L_i, L_i' are disjoint and their union is H, it is sufficient to prove the first and third assertions with equality replaced by inclusion. Let $z \in L_i$; then the H-line zw_i crosses λ_i at a point z', and $d(w_0, z') = d(z', w_i)$. By the triangle inequality

$$d(w_i, z') = d(w_0, z') > d(w_0, z) - d(z, z');$$

hence

$$d(w_i, z) = d(w_i, z') + d(z', z) > d(w_0, z).$$

That is, $L_i \subset \{z \in \mathrm{H} \mid d(z, w_0) < d(z, w_i)\}$, and the same argument works for L_i'.

We now define

$$N = \bigcap_{i=1}^{\infty} L_i .$$

Anticipating our later discussion we call N *a normal polygon with center* w_0. By definition N is a subset of H; it does not contain any points of E. In view of the above lemma we may write

$$N = \{w \in \mathrm{H} \mid d(w, w_0) < d(w, w_i) \qquad \text{for all } i > 0\}.$$

This may be expressed by saying that N consists of those points of H that are *strictly nearer* w_0. We are going to prove that N is a fundamental region for Γ.

N is not empty, for it contains the point w_0. Moreover N is H-convex. Indeed, each L_i is H-convex, therefore their intersection is also. As a consequence N is connected.

Lemma 2. A compact subset of H meets only finitely many bisectors λ_i.

Since H is a metric space under hyperbolic distance, a compact subset K is bounded; that is, there is an $R < \infty$ that is an upper bound for the distances from w_0 to points of K. Let w' be a point of λ_i lying in K; then

$$d(w_0, w_i) \leq d(w_0, w') + d(w', w_i) = 2d(w_0, w') \leq 2R.$$

That is, λ_i meets K only if w_i lies in the closed hyperbolic disk of radius $2R$ about w_0. This disk is compact and can contain only finitely many w_i, otherwise it would contain an accumulation point of the $\{w_i\}$ that would not be an ordinary point.

Lemma 3. N is open.

Let $w \in N$. A closed disk K containing a point w of N meets only finitely many λ_i, as we just proved. Since w is not on any λ_i, there is a smaller disk K that meets no λ_i. Consider a single L_i. Either K lies within L_i, or K cuts the boundary of L_i (that is, λ_i), or K lies in the exterior of L_i. The second case has been excluded, and the third case is impossible because w, as a point of N, lies within L_i. Hence $K \subset L_i$, and this is true for each i. It follows that $K \subset \bigcap_i L_i = N$.

We have shown that the normal polygon is a nonempty H-convex region. From now on we denote N by N_0 and define

$$N_i = V_i(N_0), \qquad V_i \in \Gamma.$$

From the invariance of H-distance we deduce that N_i consists of the points of H that are strictly nearer w_i. Like N_0, N_i is a nonempty H-convex region. It is called a normal polygon of Γ with center w_i. The regions N_i are permuted among themselves by the transformations of Γ, but the whole family or network of regions $\{N_i, i = 0, 1, \cdots\}$ is invariant under Γ. This network is called a *partition* or *tessellation of* H *with center* w_0. The polygons of a partition are mutually H-congruent. To a noneuclidean observer the configuration of polygons looks exactly the same when viewed from each of the points $\{w_i\}$. A picture of a famous tessellation, that of the modular group, is shown in Figure 3.

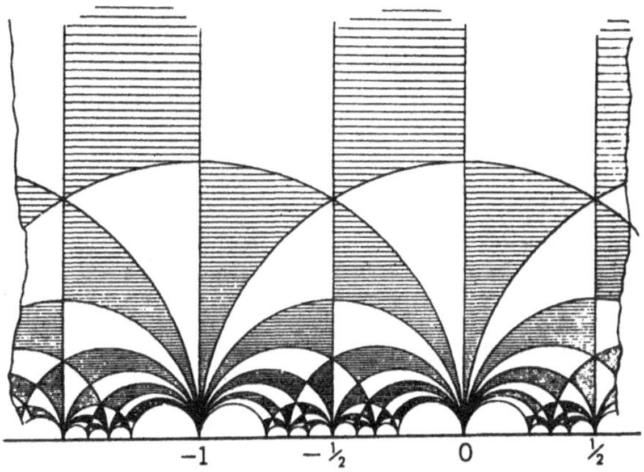

FIG. 3. The Modular Group. A shaded and unshaded region together constitute a fundamental region for the group. (Reprinted with the courtesy of B. G. Teubner, Stuttgart, Germany.)

LEMMA 4. $N_i \cap N_j = \emptyset$ for $i \neq j$.

Suppose $w \in N_i$, $w \in N_j$. Since $w \in N_i$, we have $d(w, w_i) < d(w, w_j)$. Since $w \in N_j$, we have $d(w, w_j) < d(w, w_i)$, a contradiction.

LEMMA 5. Two distinct points of N_i are inequivalent.

Suppose $\zeta_1, \zeta_2 \in N_i$ and $V\zeta_1 = \zeta_2$, $V \neq I$. Since V maps N_i on some N_j, we have $\zeta_2 \in N_j$, or $N_i \cap N_j \neq \emptyset$. This contradicts Lemma 4.

LEMMA 6. Every point of H lies in exactly one of the following sets:

$$\begin{aligned} I &= \{z \in \mathrm{H} \mid d(z, w_0) < d(z, w_i) && \text{for all } i \neq 0\} \\ B &= \{z \in \mathrm{H} \mid d(z, w_0) \leq d(z, w_i) && \text{for all } i, \\ & \qquad\qquad d(z, w_0) = d(z, w_k) && \text{for at least one } k \neq 0\} \\ E &= \{z \in \mathrm{H} \mid d(z, w_0) > d(z, w_k) && \text{for at least one } k \neq 0\} \end{aligned}$$

Moreover,

$$I = \operatorname{Int} N_0,\ B = \operatorname{Bd} N_0 \cap \mathrm{H}, \qquad E = \operatorname{Ext} N_0 \cap \mathrm{H}.$$

We have only to prove the last line. That $I = \operatorname{Int} N_0$ is obvious, for $I = N_0$ by definition and N_0 is open. Suppose $z \in B \cap \mathrm{H}$, then z belongs to every $\bar{L}_i$. On the other hand, by Lemma 1, z lies on only finitely many λ_i, say for $i = 1, \cdots, s$. Hence every sufficiently small neighborhood K of z lies in each L_i for $i > s$. The λ_i with $i = 1, \cdots, s$ divide H into a finite number of regions, one of which, say D, contains w_0 as an inner point, since w_0 lies on no λ_i. Let λ^* be the H-line joining w_0 and z. Since λ^* meets each λ_i at z, it can have no other point of intersection with λ_i in H. It follows that λ^* lies on the same side of λ_i as w_0; in other words λ^* lies in L_i. Since this is true for $i = 1, \cdots, s$, we conclude that $\Delta = K \cap D$ is in every L_i, $i \geq 1$, and is therefore in N_0. Thus K meets N_0, so $z \in \bar{N}_0$. But B is disjoint from I and therefore $z \in \operatorname{Bd} N_0$.

Conversely, a point z of H lying on Bd N_0 lies in $\bar{N}_0 = Cl \bigcap_i L_i \subset \bigcap_i \bar{L}_i$. Hence $d(z, w_0) \leq d(z, w_i)$ for all i. But $z \notin N_0$, so the conditions defining I must be violated for at least one k. Therefore $z \in B$.

Finally, E is the complement of $I \cup B$, and this concludes the proof.

LEMMA 7. $\cup_i \bar{N}_i$ covers H.

Let $w \in \mathrm{H}$. The set $\{w_i\} = \Gamma w_0$ does not accumulate at w and so contains a w_j nearest to w:

$$d(w_j, w) \leq d(w, w_i) \qquad \text{for all } i.$$

Hence with $w' = V_j^{-1}w$ we have

$$d(w_0, w') \leq d(w', w_i) \qquad \text{for all } i,$$

or, by Lemma 6, $w' \in \bar{N}_0$. Thus $w \in V_j\bar{N}_0 = \bar{N}_j$ and the lemma follows.

We can now prove the main result.

THEOREM. N_0 (and therefore N_j) is a fundamental region for Γ relative to H. Any compact subset of H is covered by finitely many polygons $\bar{N}_i$.

The proof of the first statement is contained in Lemmas 3, 5, and 7. Let K be the compact subset. Since K is certainly covered by closed normal polygons, we have to show only that K meets no more than a finite number of them. Consider $N_j = V_j N_0$, the normal polygon with center $w_j = V_j w_0$. Since w_j lies in L_j' (the half-plane complementary to L_j), it follows that N_j lies in L_j', since $z \in N_j$ only if $d(z, w_j) < d(z, w_0)$. Now $d(w_0, w_j) \to \infty$ with $j \to \infty$, and $d(w_0, \lambda_j) = \frac{1}{2} d(w_0, w_j)$; hence $d(w_0, N_j) \to \infty$. For any R, then, only a finite number of N_j intersect the H-disk D_R about w_0 of radius R. But K is compact and lies in D_R for some R.

Here is a second proof. Suppose ζ lies in $N_j \cap K$. Let ζ_0 be the image of ζ in N_0 so that $\zeta = V_j \zeta_0$. If $R < \infty$ is an upper bound for the H-distance of points in K from w_0, we have $d(w_0, \zeta) \leq R$, which implies $d(w_k, \zeta_0) \leq R$, where $w_k = V_j^{-1} w_0$. Now

$$d(w_0, w_k) \leq d(w_0, \zeta_0) + d(\zeta_0, w_k) \leq d(w_0, \zeta_0) + R.$$

To get a bound on $d(w_0, \zeta_0)$ recall that for $\omega \in N_0$ we have $d(\omega, w_0) < d(\omega, w_i)$ for all $i > 0$. In particular, $d(w_0, \zeta_0) < d(\zeta_0, w_k) \leq R$. Hence

$$d(w_0, w_k) < 2R.$$

Thus the only polygons that meet K are those whose centers lie in an H-disk of radius $2R$, and there are only finitely many of these.

Exercise 1. Construct normal polygons for an elliptic, parabolic, and hyperbolic cyclic group.

Exercise 2. Show that

$$N = \{z \in \mathrm{H} \mid d(z, w_0) < d(Vz, w_0) \quad \text{for } V \in \Gamma,\ V \neq \mathrm{I}\}.$$

4D. We are now going to study the boundary of a normal polygon, say N_0 with center w_0. According to 4C, Lemma 6,

$$\begin{aligned} \operatorname{Bd} N_0 \cap \mathrm{H} = \{w \in \mathrm{H} \mid d(w, w_0) &\leq d(w, w_i) \quad \text{for all } i \text{ and} \\ d(w, w_0) &= d(w, w_j) \quad \text{for at least one } j\}. \end{aligned} \tag{16}$$

That is, the part of the boundary of N_0 lying in H consists of those points that lie on a finite number of bisectors and lie on the w_0-side of the remaining bisectors.

Suppose $w \in \mathrm{H}$ satisfies these conditions and lies on exactly one bisector λ_j. In a sufficiently small neighborhood of w there is another boundary point w' lying on λ_j. Because $\bar{N}_0$ is still H-convex, the whole arc ww' belongs to the boundary of N_0. Thus there is a largest arc of λ_j contained in Bd N_0, and we call this arc, with or without its end points, a *side*. Since only a finite number of bisectors meet any H-disk centered at w_0, say, we see that N_0 has at most a denumerable number of sides on its boundary. The condition that w be an inner point of a side is therefore

$$d(w, w_j) = d(w, w_0) < d(w, w_i) \qquad \text{for exactly one } j \text{ and all } i \neq j. \tag{17}$$

The remaining boundary points of N_0 in H must lie on more than one bisector; they are called *ordinary vertices*. The ordinary vertices lie isolated in H, since only a finite number of bisectors cross a neighborhood of a vertex and two bisectors intersect in at most one point of H. Let v be an ordinary vertex and let K be a disk containing v, containing no other vertex, and meeting no bisectors except those that pass through v. The finite number of bisectors that pass through v, say $\lambda_1, \cdots, \lambda_s$, divide K into $2s$ sectors, and as we saw in the proof of 4C, Lemma 6, one of them, say Δ, lies entirely in N_0. Let λ_1', λ_2' be the portions of the bisectors bounding Δ. Every neighborhood of a point on λ_1' meets Δ, therefore meets N_0. It follows that λ_1' and λ_2' are part of the boundary of N_0. Therefore λ_1', λ_2' are parts of sides of N_0. But not more than two sides can intersect in v, otherwise the convexity of N_0 would be violated. *An ordinary vertex is the intersection of exactly two sides of* N_0. Conversely, the intersection of two sides of N_0, if it lies in H, is clearly a boundary point, and therefore an ordinary vertex. We may now redefine an ordinary vertex of N_0 to be a point of H lying on exactly two sides of N_0.

To summarize: the part of the boundary of N_0 contained in H consists of a countable number of sides and ordinary vertices.

We now consider the boundary points of N_0, if any, that lie on E. In what follows all sets will be considered to be subsets of $\bar{\mathrm{H}} = \mathrm{H} \cup \mathrm{E}$. If α is such a point, then either

(1) α lies on exactly two sides of N_0,

(2) α lies on exactly one side of N_0,

(3) α does not lie on any side of N_0.

This exhausts the possibilities since α cannot lie on more than two sides without violating the convexity of N_0.

In case (1) let us assume that when one of the sides s is described from a point H to α, N_0 lies to the left. Then the other side t must be placed so that when it is described from α to a point in H the region N_0 lies to its left, because of convexity. Since s and t are both orthogonal to E, they are tangent at α. They enclose a sector that is part of N_0; for, again by convexity, a half-ray joining w_0 to any point of either s or t lies entirely in N_0.

Before treating case (2) let us note the following general result on convex regions, valid for H-convexity as well as for euclidean convexity.

LEMMA. A convex region is a star with respect to any interior point.

We say a region R is a star with respect to the interior point c if every half-line issuing from c contains exactly one boundary point of R. Suppose R is convex. Then R contains, with every point $\zeta \in R$ on the half-line l from c, the entire segment $c\zeta$. There exists a maximal open segment $c\omega$ on l lying in R, and ω is clearly a boundary point of R. Suppose ω' is a point on l on the other side of ω from c. By convexity no point on $\omega\omega'$ is a point of R. Thus ω' cannot be a boundary point of R, and the result is proved.

Case (2) embraces two possibilities. It may be that there is no side of N_0 other than s meeting a certain neighborhood D of α. According to the lemma the open H-line l joining w_0 to α lies entirely in N_0, for α is a boundary point of N_0. Let us consider triangles Δ with vertices w_0, α, β, where $\beta > \alpha$ lies on E and $l' = w_0\beta$ intersects D. Write the disjoint union

$$\Delta = K_1 \cup K_2,$$

where $K_1 = \Delta \cap D$. Because of the present assumption no sides of N_0 meet K_1. Since $\bar{K}_2$ is compact, only a finite number of sides intrude into K_2. None of

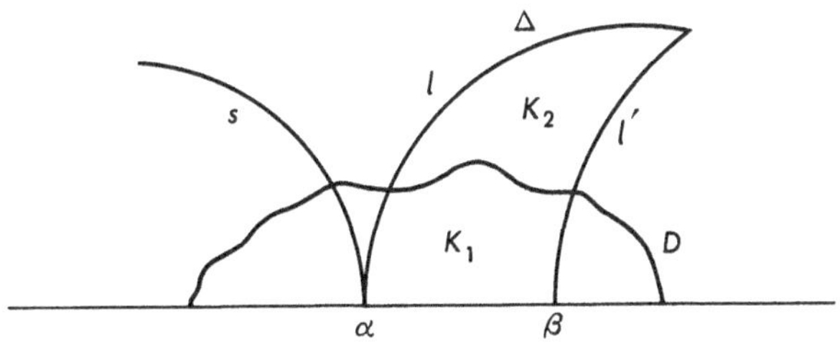

FIG. 4.

these sides, however, intersects l; for the point of intersection would be a boundary point of N_0. Hence there is a β such that the triangle Δ is free of sides of N_0. Let Δ^* be the largest such triangle, obtained by rotating l' as far as possible to the right, and let γ be the end point of l' on E. Here γ can be ∞. The interior of Δ^* belongs to N_0 and the interval $(\alpha, \gamma) = f$ is part of the boundary of N_0. We call f a *free side.*†

If a side s terminates in E with N_0 lying to its *right*, we can, of course, follow the same procedure. Thus there may be a semi-infinite free side $(-\infty, \delta)$, where δ is the end point of s on E. If there are free sides $f_1 = (\alpha_1, \infty)$ and $f_2 = (-\infty, \alpha_2)$, where, of course, $\alpha_2 < \alpha_1$, we shall combine them and call the union of f_1 and f_2 a single free side.

As the second possibility under case (2) it may happen that s is the only side on which α lies, but every neighborhood of α meets sides of N_0 (from the right). This involves an infinite number of sides for N_0 and can actually happen.‡

If γ is finite, there may be a side of N_0 issuing from γ, or there may be no such side but infinitely many sides intersecting every neighborhood of γ. There can be no free side beginning at γ and proceeding to the right, since by definition a free side is maximal.

† The word *side* shall be reserved for boundary arcs or lines in H.

‡ For examples of such fundamental regions see Ford, pp. 57–58; Lehner, pp. 124–125.

We call an end point of a free side, or an end point of a side lying on E, a *real vertex*.

In case (3) the point α may be an inner point of a free side. If not, every neighborhood of α must meet infinitely many sides from both left and right. For an H-line l drawn from w_0 to α lies wholly in N_0, but when rotated in either direction l meets sides of N_0, otherwise α would lie on a free side.

THEOREM. The part of the boundary of N_0 that lies in H consists of a countable number of sides and ordinary vertices. The part of the boundary that lies in E is either empty, or consists of a countable number of real vertices and free sides, and of points whose every neighborhood meets infinitely many sides. When N_0 has only a finite number of sides, the last class is empty. The inner points of a free side are ordinary points of Γ.

Only the last statement requires proof. Let α be an inner point of a free side f and assume $\alpha \in$ L. According to Theorem 3A there is a point $\beta \in f$, $\beta \neq \alpha$, such that the set $\Gamma\beta$ accumulates at α. Hence, since α is an inner point of f, there are infinitely many equivalent points lying inside an open subinterval of f. But we can easily prove that *two distinct inner points of a free side are never equivalent under Γ*. In fact, let ζ_1, and $\zeta_2 = V\zeta_1$, $V \in \Gamma$, $V \neq$ I, be the points, and let A_1 be a plane neighborhood of ζ_1 such that $A_1 \cap \text{H} \subset N_0$. This is possible since ζ_1 is an inner point of f and therefore a boundary point of N_0. Then $VA_1 = A_2$ is a neighborhood of ζ_2; hence A_2 meets N_0. If z is a point of both A_2 and N_0, then $V^{-1}z \in A_1 \cap N_0$; that is, both z and $V^{-1}z$ lie in N_0. This is a contradiction, which concludes the proof.

We note that Γ is horocyclic (3D) if and only if the normal polygon has no free sides.

4E. Let us now attempt a more global survey of the boundary of N_0, which we denote by B. We shall define a countable family of essentially disjoint connected subsets of B called *boundary components*.

Let P be an inner point of a side s_1 of N_0. We follow s_1 in the positive direction (N_0 lies to the left) and encounter at its end a vertex v_1, either ordinary or real. In the first case there is a unique side s_2 issuing from v_1; in the second case, either a side s_2 or a free side f_2 issues from v_1, or no side at all. Disregarding the last possibility we continue along s_2 or f_2, which terminates in a vertex v_2, etcetera. The continuation of this procedure leads to three possibilities:

(1) After encountering a finite number of sides or free sides or both, we return to P.

(2) We encounter a finite number of sides in H, the last of which, s_k, terminates in a point v_k on E from which no side other than s_k issues.

(3) We can continue the process indefinitely without ever returning to P or arriving at a point of E.

Next, we start at P and follow B in the *negative* direction. Then each of the three possibilities can occur, except that (1) occurs if and only if it occurred in the course of the positive description of B.

Let

$$\cdots, s_{-j}, v_{-j}, \cdots, s_{-1}, v_{-1}, s_0, v_0, s_1, v_1, \cdots, s_j, v_j, \cdots \tag{18}$$

be the sequence of sides, free sides, and vertices in the order in which they are encountered as we describe B from P in both directions. In case (1) the sequence breaks off at $v_n (= v_0)$ and at $v_{-n} = v_0$, for some n. If we describe B positively, (18) terminates with v_k in case (2) and is infinite in the positive direction in case (3). Thus there are four possible types of sequences (18).

In case (18) is infinite in the positive direction, v_j converges to a point v_∞ on E as $j \to \infty$; similarly $v_j \to v_{-\infty} \in \mathrm{E}$ as $j \to -\infty$ if (18) is infinite in the negative direction. We adjoin v_∞ and $v_{-\infty}$, when they occur, to the sequence (18). We now define the *boundary component of B determined by P* to be the union of the points lying on the sides and vertices of the sequence (18). Clearly a boundary component is independent of the point P and is uniquely determined by any of its points except the end points $v_{-\infty}$, v_∞ (if they should be present). A boundary component is a closed Jordan arc; for parameter we can use the angle made with a fixed line by an H-line connecting the center w_0 to a point of the boundary component. Distinct boundary components intersect at most in a common end point.

Let $\{B_\alpha\}$ be the distinct boundary components of B. Draw H-lines from w_0 to the end points of B_α, forming a hyperbolic convex region R_α. By the convexity of N_0, all of R_α lies in N_0. Since the angle subtended by the sides of R_α at w_0 is positive, *there are at most denumerably many boundary components.*

Exercise 1. The complement of $\cup_1^\infty B_\alpha$ in B is nowhere dense in B.

Let us now consider more particularly the case in which the boundary of N_0 consists of a finite number of sides and free sides. Then all boundary components must be of type (1); for types (2) and (3) necessarily involve an infinite number of sides. Now $\cup B_\alpha = B$; for every boundary point q of N_0 by assumption lies on a closed side or free side and so determines a boundary component and, in fact, a unique boundary component.

We join two successive vertices on B_1 to w_0, forming triangles Δ_k, $k = 1, \cdots, m$, which by the convexity of N_0 lie wholly in N_0. Let P_1 be the union of the $\bar{\Delta}_k$; we define P_α similarly. Then P_1 intersects some P_α, $\alpha \neq 1$, otherwise $\bar{N}_0$ is not connected. If $p \in P_1 \cap P_\alpha$, the point q which is the intersection of the H-line $w_0 p$ with B is a point lying in $B_1 \cap B_\alpha$. But as we remarked above, q determines a unique boundary component. Hence there is only one boundary component B_1 and $B_1 = B$.

It follows that $P_1 = \bar{N}_0$. Otherwise there would be a point p in $\bar{N}_0$ such that the H-line through w_0 and p meets B in a point that does not belong to B_1.

Thus N_0 is the union of a finite number of triangles each of which has w_0 as a vertex and has a side issuing from w_0 in common with exactly two other triangles. Hence N_0 is a simply connected *polygon*. We have proved the following

THEOREM. When N_0 has a finite number of sides and free sides, it is a convex H-polygon in the closed upper half-plane $\bar{\mathrm{H}}$.

4F. Conjugate Sides. Suppose s is a side of N_0; by 4D, (17), the point z is an inner point of s if and only if there is a $j \neq 0$ such that

$$d(z, w_j) = d(w_0, z) < d(z, w_i) \qquad \text{for } i \neq 0, i \neq j.$$

Applying $V_j^{-1} = V_k$, we get

$$d(w_0, V_k z) = d(w_k, V_k z) < d(V_k z, w_l) \qquad \text{for } l \neq 0, l \neq k,$$

since V_k maps $\{w_i\}$ on itself. This says $V_k z$ is a boundary point of N_0 and in fact an inner point of a side s' of N_0. Thus $V_k s \subset s'$.

The inner points w of the side s' are characterized by the condition

$$d(w_n, w) = d(w, w_0) < d(w, w_i) \qquad \text{for } i \neq 0, i \neq n,$$

where $n > 0$ is a certain integer. Since for some points on s' (namely, images of points on s) we have $n = k$, this is true for all points on s', and the condition

$$d(w_k, w) = d(w, w_0) < d(w, w_i) \qquad \text{for } i \neq 0, i \neq k,$$

holds for all inner points w of s'. Hence with $z = V_k^{-1} w$,

$$d(w_0, z) = d(z, w_j) < d(z, w_i) \qquad \text{for } i \neq 0, i \neq j.$$

This says $V_k^{-1} w$ is an inner point of s and so $V_k^{-1} s' \subset s$. Therefore V_k maps s on s'. Since, as we have seen, inner points of s go into inner points of s' and, conversely, end points go into end points, it follows that V_k carries the whole open (closed) side s onto the whole open (closed) side s'. *Every side s of N_0 is equivalent to a side s' of N_0 by a $V \in \Gamma$, $V \neq I$.*

We observe that equivalent points lying on sides of N_0 are equidistant from w_0. Indeed, if z and Vz are two such points, we have

$$d(w_0, z) \leq d(z, w_i) \leq d(V_i^{-1} z, w_0) \qquad \text{for all } i; \tag{19}$$

in particular ($V_i^{-1} = V$), $d(w_0, z) \leq d(w_0, Vz)$. Reversing the roles of z and Vz we get $d(w_0, Vz) \leq d(w_0, z)$, or $d(w_0, z) = d(w_0, Vz)$.

Could s be equivalent to more than one side of N_0? To show this is not possible, we shall prove that no *inner* point of s is equivalent to more than one point on the boundary of N_0.

Let z be an inner point of a side s, satisfying

$$d(w_0, z) = d(z, w_j) \qquad \text{for some } j. \tag{20}$$

Let $z' = Vz$ lie in Bd N_0. Then z' lies on a side s' and satisfies

$$d(w_0, z') = d(z', w_k)$$

for some k. Let $Vw_m = w_0$; we have

$$d(z, w_m) = d(z', w_0) = d(z', w_k) = d(z, w_0),$$

since z and z', as equivalent boundary points, are equidistant from w_0. Using (20),

$$d(z, w_m) = d(z, w_0) = d(z, w_j).$$

That is, z lies on *two* bisectors, those determined by w_m and w_j. Since z is not a vertex, we must have $w_m = w_0$ or $w_m = w_j$. The first implies $V^{-1}w_0 = w_0$, but w_0 is not a fixed point. Hence $w_m = w_j$; that is, $V_j^{-1}V^{-1}w_0 = w_0$. For the same reason this implies $V = V_j^{-1}$. The point z' is $V_j^{-1}z$ and is unique, being determined by the fact that z is an inner point of s.

We shall say two sides of N_0 are *conjugate* if there is an element of Γ, not the identity, mapping one on the other. We have proved:

THEOREM. The sides of N_0 are conjugate in pairs.

This theorem applies only to sides lying in H; it is false for free sides. As we have seen, an inner point of a free side is not equivalent to any point of $\bar{N}_0$ except itself (see end of 4D).

Nothing we have said so far rules out the possibility that two conjugate sides are identical. If $Vs = s$ with $V \neq I$ and s is the H-segment ab, then either a and b are individually fixed by V or else they are interchanged. The first case is impossible. In fact, no transformation of Ω_R has two fixed points in H nor does it have one on E and one in H. If a and b are both on E, V must be hyperbolic; however we shall show later (4I) that the fixed point of a hyperbolic transformation never lies in the closure of a normal polygon. We must therefore assume that $Va = b$, $Vb = a$. Thus a is a fixed point of V^2, and b is also. By the previous reasoning $V^2 = I$, so V is elliptic of order 2.

Let ξ be the fixed point of V in H and let α be the H-mid-point of ab. We have $d(a, \alpha) = d(b, \alpha)$ by construction. But also $d(a, V\alpha) = d(Va, V^2\alpha) = d(b, \alpha)$, and similarly $d(b, V\alpha) = d(a, \alpha)$. Hence $d(a, V\alpha) = d(b, V\alpha)$. Since V maps s on itself, $V\alpha$ lies on s and must therefore coincide with α. In other words α is a fixed point of V and so $\alpha = \xi$.

We conclude that a side s coincides with its conjugate side if and only if there is an elliptic element $V \in \Gamma$ of order 2 that has a fixed point α coinciding with the H-mid-point of s. Thus α divides s into two segments of equal length and V maps one segment on the other. We shall agree in this situation to call α a vertex. *The fixed point α of an elliptic element V of order 2 will be considered to be a vertex, and the two pieces of the original side s containing α that are interchanged by V will be considered to be distinct sides meeting at α.*

According to this convention a side can be mapped on itself only by the identity.

Exercise 1. Conjugate sides are of equal H-length.

Exercise 2. Let $(s_1, s_1' = T_1 s_1)$, $(s_2, s_2' = T_2 s_2)$ be distinct pairs of conjugate sides. Then $T_1 \neq T_2$.

4G. Incidence Pattern. We shall investigate next how the normal polygons fit together in the tessellation of H. Suppose N_j and N_k have a common boundary ω. Since $\bar{N}_j$ and N_k themselves do not intersect, it is clear that ω is simply $\bar{N}_j \cap \bar{N}_k$. (All closures are taken with respect to the closed upper half-plane.) Hence ω is convex since $\bar{N}_j$ and $\bar{N}_k$ are convex; it is a closed H-convex subset of Bd N_j (and of Bd N_k). As such it can be a single point, but as soon as it contains two points, it contains the whole H-segment joining them. This shows at once that the part of ω lying on E is empty, a single point, or two points.

If ω contains an H-segment, this H-segment contains inner points of a side s of N_j. Hence the whole closed side $\bar{s}$ belongs to $\bar{N}_j$ as well as to $\bar{N}_k$ and therefore to ω. Thus if ω is not a single point, it contains a common side of N_j and N_k. Can ω contain other points not on a common side? Usually different sides of N_j lie on different bisectors, and when this is the case ω cannot contain inner points of different sides, since then it would not be convex. However, there is one case in which two sides lie on one bisector and are separated by an ordinary vertex, as we saw at the end of 4F. In this exceptional case ω can be the union of two closed sides common to N_j and N_k. In all cases, however, the common boundary of two normal polygons is a closed H-segment, which may reduce to a point, or is empty.

Next, suppose ω reduces to a single point α of H. Then α is not an inner point of a side, otherwise ω would contain the whole side. Hence α is a vertex and lies on a finite number of bisectors. We draw an open disk K about α small enough so that it contains no other vertex and meets no bisectors other than those passing through α. K is partitioned by the bisectors into a finite number of open triangles Δ_i, $i = 1, 2, \cdots, s$. A point z_1 of Δ_1 lies on no bisector and is therefore an inner point of some normal polygon, say N_1. Suppose $z_2 \in \Delta_1$ belongs to a different normal polygon N_m. Because of the connectedness of Δ_1, we can join z_1 to z_2 by an arc γ that stays inside Δ_1 and hence meets no bisector. But γ must leave N_1 by a boundary point of N_1, which contradicts the fact that γ meets no bisector. Hence $\Delta_1 \subset N_1$.

We apply the same process to the remaining Δ's. Each Δ_i lies in some normal polygon N_i and these normal polygons are all different. If, for example, Δ_m and Δ_n would both be subsets of N_m, an H-line joining a point of Δ_m to one of Δ_n would cross one of the bisectors bounding Δ_m and in doing so violate the convexity of N_m. Now α is obviously a boundary point of each N_i. But α is not an inner point of any side, hence α is a vertex of N_i, $i = 1, \cdots, s$.

THEOREM. The common boundary ω of two normal polygons, if not empty, is a point or a closed H-segment. If ω is a point lying in H, it is the common vertex of a finite number of normal polygons and its immediate neighborhood is covered by the closures of these polygons. If ω is an H-segment, it is either a closed side or else the union of two closed sides that meet at the fixed point of an elliptic transformation of order 2.

If ω lies on E, there can be infinitely many normal polygons with vertex ω, as we shall see in 4I. Or ω can be the common end point of two free sides lying in different normal polygons.

Exercise 1. Under a transformation V of Γ vertices of N_i go into vertices of VN_i and sides of N_i go into sides of VN_i.

4H. Cycles. The relation of Γ-equivalence partitions the ordinary vertices of N_0 into equivalence classes called *ordinary cycles*. An ordinary cycle of N_0 is therefore a set consisting of an ordinary vertex of N_0 together with all other vertices of N_0 equivalent to it. Obviously an ordinary cycle lies wholly in H.

THEOREM 1. An ordinary cycle contains only a finite number of vertices.

This theorem is trivial only if N_0 has a finite number of sides, which we are not assuming. Let $C = \{z_1, z_2, \cdots\}$ be an ordinary cycle. Let $z_1 = T_j z_j$, $j = 1, 2, \cdots$. T_j carries N_0 into a normal polygon N_j which has z_1 as a vertex, and these polygons are distinct. If C is infinite, there are infinitely many normal polygons meeting at z_1, which contradicts Theorem 4G.

If one vertex in C is a fixed point of an element of Γ (or, as we say, a fixed point of Γ), the same is true of every vertex, for TET^{-1} fixes Tz if E fixes z. Furthermore, a transformation fixing an ordinary point is necessarily elliptic. We therefore classify an ordinary cycle as *elliptic* or *accidental*, according as all or none of its vertices are fixed points, and the corresponding vertices are called elliptic vertices or accidental vertices. (The term *accidental* arises from the fact that the distribution of accidental cycles in N_0 depends on the choice of center w_0, whereas the number of elliptic cycles is always the same, as we shall see presently.) The stabilizer Γ_v of an elliptic vertex v is a finite cyclic group (2H) whose order is called the order of the vertex. Since $\Gamma_{Tv} = T\Gamma_v T^{-1}$, the order of all vertices of an elliptic cycle is the same. This integer is also called the order of the cycle.

The sides of N_0 that meet at a vertex v are circular arcs and form two angles if v is ordinary. The measure of that angle which bounds a portion of N_0 will be termed *the angle at v in N_0*. When v is on E, we say the angle is zero if v is the intersection of two sides and $\pi/2$ if v is the intersection of a side and free side.

By definition every elliptic cycle is made up of fixed points of elliptic elements of Γ. Conversely, we have the following:

THEOREM 2. Every elliptic fixed point is a point of an elliptic cycle of some normal polygon.

Let the elliptic fixed point α be fixed by an element E of period l. An application of E amounts to a noneuclidean rotation about α through an angle $2\pi/l$. Now α is a point of H and is therefore found in the closure of some normal polygon N_j. Clearly α is no interior point of N_j, for a sufficiently small neighborhood would lie in N_j and contain l points equivalent under E. If α is an inner point of a side s of N_j and $l > 2$, we can still find two equivalent points on the same side of s and therefore in N_j. If $l = 2$, we have the situation discussed at the end of 4F, where we agreed to call α a vertex.

In all cases, then, α is a vertex of N_j. Since it is an elliptic fixed point, it is a member of an elliptic cycle of N_j.

THEOREM 3. The sum of the angles at the vertices of an ordinary cycle of N_0 is $2\pi/l$ if and only if the cycle is elliptic of order l, and is 2π if and only if the cycle is accidental.

Let $C = \{z_1, z_2, \cdots, z_s\}$ be an elliptic cycle of order $l \geq 2$. Let E be a generator of the stabilizer Γ_{z_1}, which we recall is cyclic, and let $T_j \in \Gamma$ be fixed transformations such that $T_j z_j = z_1, j = 1, 2, \cdots, s$. The elements of Γ that map z_j on z_1 are precisely

$$A_j = \{E^k T_j \mid 0 \leq k < l\}.$$

Indeed, from $Vz_j = z_1$ we deduce that VT_j^{-1} fixes z_1 and is therefore a power of E.

The immediate neighborhood of z_1 is completely covered by the closures of normal polygons, each of which has z_1 as a vertex (Theorem 4G). Let N_1 be one of these polygons and let $W(N_0) = N_1$, $W \in \Gamma$. Then $W^{-1}z_1$ is a vertex of N_0 and this vertex can only be a member of C, say z_j. Hence $W \in A_j$. Conversely, each W in A_j maps N_0 onto a normal polygon that has a vertex at z_1. The neighborhood of z_1 is made up of the closures of sl normal polygons, the images of N_0 by the elements $\bigcup_1^s A_j$.

Denote by θ_j the angle at z_j in N_0. The angle θ_{kj}, defined as the angle at z_1 in the normal polygon

$$N_{kj} = E^k T_j(N_0), \qquad k = 0, 1, \cdots, l-1$$

is equal to θ_j because of the conformality of the transformations of Γ. Hence

$$2\pi = \sum_{j=1}^{s} \sum_{k=0}^{l-1} \theta_{kj} = l \sum_{j=1}^{s} \theta_j ,$$

as promised. The situation is illustrated for the case $s = 2$, $l = 3$.

If C is an accidental cycle, there is a unique T_j that carries z_j to z_1, $j = 1, \cdots, s$. The s normal polygons $T_j N_0, j = 1, \cdots, s$ make up the neighborhood of z_1. The angle at z_1 in $T_j N_0$ is equal to the angle at z_j in N_0. This prove- the direct part of the theorem.

For the converse note that if C is a cycle with angle sum $2\pi/l$, $l > 1$, then by what we have just proved, C cannot be elliptic of order $m \neq l$ nor can it be accidental. If the angle sum equals 2π, C cannot be elliptic of any order and so must be accidental. This concludes the proof.

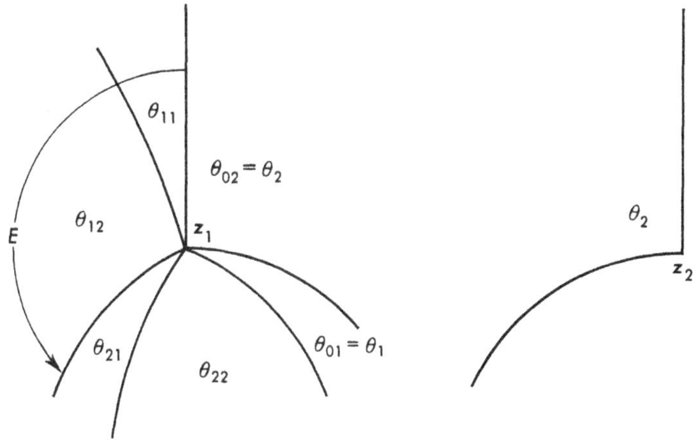

FIG. 5.

Let α be an elliptic fixed point of order l and let K be the interior of a fixed circle of E, a generator of Γ_α. For a fixed j the regions $E^k T_j N_0$, $k = 0, 1, \cdots, l-1$, are equally spaced around α, each being obtained from the preceding one by a rotation through the angle $2\pi/l$. We can select $k = k_j$ (depending on T_j) so that the regions $\{E^{k_j} T_j N_0, j = 1, 2, \cdots, s\}$ are adjacent. We call

$$S_\alpha = \left\{ \bigcup_{j=1}^{s} E^{k_j} T_j \bar{N}_0 \right\} \cap K$$

an *elliptic sector at* α; it subtends an angle of $2\pi/l$ and the sectors $\bar{S}_\alpha$, $E\bar{S}_\alpha$, $\cdots$, $E^{l-1}\bar{S}_\alpha$ make up the neighborhood of α.

Exercise 1. The points of an ordinary cycle are H-equidistant from the center w_0.

Exercise 2. By suitable choice of the center w_0 we can ensure that each ordinary cycle of N_0 consists of a single vertex.
[Use the above Exercise.]

41. We next consider cycles lying on E. If one point of a cycle is on E, the entire cycle is likewise. Let $p_1 \in E$ be a vertex where two sides meet and let $\{p_1, p_2, \cdots\}$ be the complete set of points of $\bar{N}_0$ equivalent to p_1. Two sides of N_0 meet at each p_i. If we think of the boundary of N_0 described in the positive sense, each side will have an initial point and a terminal point. When a side s is mapped onto its conjugate side s' by V, the initial point of s goes into the terminal point of s', because V maps N_0 outside its boundary and orientation is preserved in a conformal mapping. Thus the initial point of s and the terminal point of s' are equivalent, likewise the terminal point of s and the initial point of s'.

We now present a practical procedure for determining all the points of a cycle if one of them is known. This method works for all cycles—those in H as well as those on E.[3]

Suppose s_1 is a side beginning at p_1. The conjugate of s_1 is a side s_1' ending at a point that is equivalent to p_1 and hence belongs to the cycle; call this point p_2. If there is a side s_2 beginning at p_2, its conjugate side s_2' ends at p_3, say. It may happen that after t steps the side s_t' ends at p_1.[4] (An example of this situation in which $t = 3$ is shown in Figure 6.) *Then we say that* $\{p_1, p_2, \cdots, p_t\}$ *is a parabolic cycle of* N_0 *and each* p_i *is called a parabolic vertex.*[†]

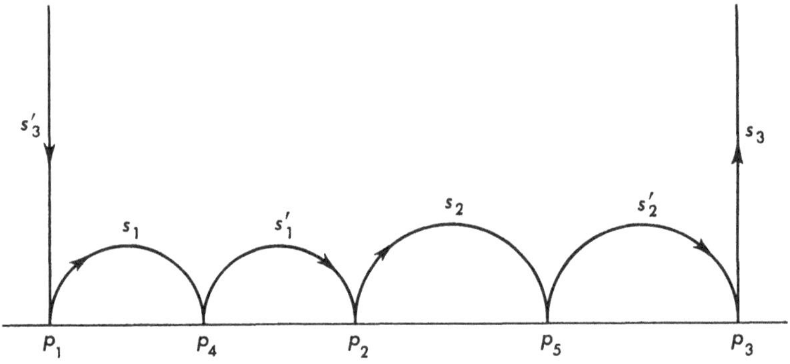

FIG. 6.

Let $W_j \in \Gamma$ be such that

$$W_j p_j = p_{j+1}, \qquad j = 1, \cdots, t-1; \qquad W_t p_t = p_1.$$

The transformation

$$P = W_t W_{t-1} \cdots W_2 W_1 \tag{21}$$

maps p_1 on itself and so is either parabolic, hyperbolic, or the identity.

[†] In Figure 6 the cycle determined by p_1 is $\{p_1, p_2, p_3\}$. The remaining cycles each consist of a single vertex and are $\{p_4\}$, $\{p_5\}$, and $\{\infty\}$.

THEOREM 1. A vertex of N_0 lying on E is never the fixed point of a hyperbolic element of Γ.

Let $T \in \Gamma$ be hyperbolic with fixed points ξ_1 and ξ_2, let w_0 be the center of N_0, and let K be the fixed circle of T that passes through w_0. The images of w_0 by powers of T lie on K. If w_1, w_2 are the nearest of these images to w_0, the region H bounded by the bisectors λ_1, λ_2 of w_0w_1, w_0w_2 constitutes a normal

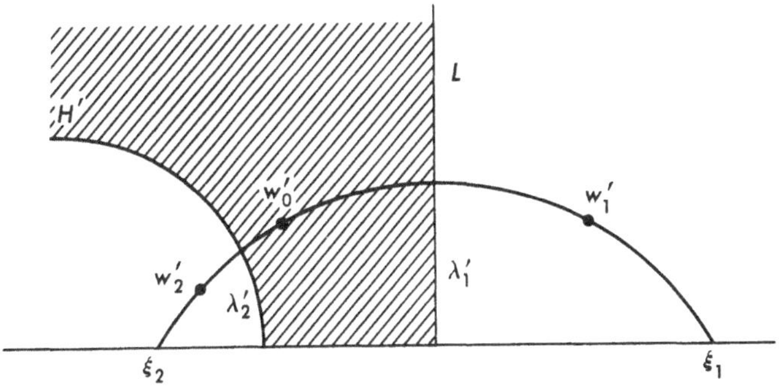

FIG. 7.

polygon for the cyclic group $\{T\}$. Now N_0 surely lies within H, for the points of N_0 have to satisfy the same inequalities as those of H and some additional ones besides. But H does not have either fixed point of T on its boundary. To see this, map H on itself by a transformation of Ω_R that preserves ξ_1 and ξ_2 and moves w_0 and w_1 to new positions w_0', w_1' symmetric with respect to L, the euclidean perpendicular bisector of the segment $\xi_1\xi_2$ in E. Then L is also the H-bisector of the H-segment $w_0'w_1'$; that is, in the normal polygon H' of the transformed group $A\{T\}A^{-1}$, $L = \lambda_1'$. This shows that H' does not have ξ_1 on its boundary, hence H does not have ξ_1 on its boundary. A similar argument proves that ξ_2 is not on Bd H, and the theorem follows.

The theorem shows that the transformation P of (21) is not hyperbolic. To show it is not the identity, we consider the arrangement of those normal polygons that have a vertex at p_1. Now W_t carries p_t to p_1 and carries N_0 into a normal polygon N_t having a vertex at p_1. Since W_t maps s_t on s_t', it is seen that N_t must have the side s_t' in common with N_0. Let s^*_{t-1} be the other side of N_t issuing from p_1: $W_ts_{t-1}' = s^*_{t-1}$. Next W_tW_{t-1} carries p_{t-1} to p_1 and carries N_0 to N_{t-1}. Since $W_tW_{t-1}s_{t-1} = W_ts_{t-1}' = s^*_{t-1}$, it is clear that N_{t-1} abuts N_t along s^*_{t-1}. Continuing in this way we obtain the regions $N_0, N_t, N_{t-1}, \cdots, N_1$ in counterclockwise order around p_1. Since $N_1 = P(N_0)$, P is not the identity.

It follows that the transformation P of (21) is *parabolic*. Thus p_1 is the fixed point of a parabolic element of Γ. Since p_i is fixed by $T_iPT_i^{-1}$ with

$T_i = W_{i-1} \cdots W_2 W_1, i = 1, 2, \cdots, t$, each parabolic vertex is the fixed point of a parabolic transformation of Γ.

The transformation P conjugates the extreme sides of the block of t polygons mentioned above. The intersection of the closures of these normal polygons with the interior K of a fixed circle of P is called a parabolic sector T_{p_1}. It is easy to verify that

$$K = \bigcup_{-\infty}^{\infty} P^m \mathrm{T}_{p_1}.$$

Let N_0 have a finite number of sides and no free sides. N_0 is a polygon, and each of its vertices, real or ordinary, is the common end point of two sides. If α is a real vertex, it necessarily determines a finite cycle. Hence α is a parabolic vertex.[5]

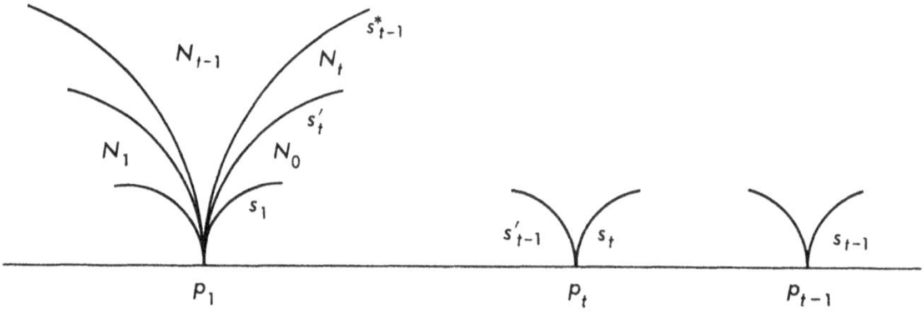

FIG. 8.

THEOREM. Each vertex p of a parabolic cycle of N_0 is the fixed point of a parabolic transformation P of Γ given by (21), and P maps one of the sides of a parabolic sector at p on the other. If N_0 has a finite number of sides and no free sides, $\bar{N}_0$ intersects E in a finite number of parabolic vertices. P generates the stabilizer Γ_p.

The last statement requires proof. Since Γ_p is cyclic (2H), it has a generator P_0. Either P_0 or P_0^{-1} maps N_0 on a normal polygon with vertex p that lies counterclockwise from it; say it is P_0. Thus P and P_0 rotate N_0 in the same sense and we have $P = P_0^m$ for some $m \geqslant 1$. It follows that $P_0(N_0) = N_j$ for a j in $1 \leq j \leq t$. Then $P_0 = W_t W_{t-1} \cdots W_j$, since the right member also maps N_0 on N_j. For $j > 1$ the right member is not in Γ_p, since its inverse does not fix p. Hence $j = 1$; that is, $P_0 = P$.

4J. We now consider the converse question: is the fixed point of a parabolic element of Γ always a parabolic vertex?

If p is the fixed point, we determine a real linear transformation that sends p to ∞ and w_0 to i. The transformed group, which we still denote by Γ, has a normal polygon with center i and contains parabolic elements fixing ∞ (translations). What we must determine is whether ∞ is a parabolic vertex of Γ.

The subgroup Γ_∞ is a parabolic cyclic group with generator

$$T = \begin{pmatrix} 1 & \lambda \\ 0 & 1 \end{pmatrix}, \qquad \lambda > 0.$$

Thus if $V = (a\, b \mid c\, d) \in \Gamma$ fixes ∞, $V = T^m$ for some integer m.

Consider the images of i under $\Gamma = \{V_n\}$:

$$V_n i = \cdots + \frac{i}{c_n^2 + d_n^2}. \tag{22}$$

If $c_n = 0$ then $d_n = \pm 1$, as we have just remarked, and Im $V_n i = 1$.

THEOREM 1. If Γ contains translations, there is no sequence $V_n \in \Gamma$ such that $c_n \to \gamma$ (finite) with distinct $\{c_n\}$. In particular, there is a constant $\tilde{c} > 0$ such that for all $V = (a\, b \mid c\, d)$ in Γ we have

$$c = 0 \qquad \text{or} \qquad |c| \geq \tilde{c}.$$

Suppose $c_n \to \gamma$, $\{c_n\}$ distinct. Let $T = \begin{pmatrix} 1 & \lambda \\ 0 & 1 \end{pmatrix}$, $\lambda > 0$, generate Γ_∞. Then

$$T^{p_n} V_n T^{q_n} = \begin{pmatrix} 1 & p_n\lambda \\ 0 & 1 \end{pmatrix} \begin{pmatrix} a_n & b_n \\ c_n & d_n \end{pmatrix} \begin{pmatrix} 1 & q_n\lambda \\ 0 & 1 \end{pmatrix}$$
$$= \begin{pmatrix} a_n + p_n c_n \lambda & \\ c_n & d_n + q_n c_n \lambda \end{pmatrix} = \begin{pmatrix} a_n' & b_n' \\ c_n & d_n' \end{pmatrix}$$

belongs to Γ. We may assume $c_n \neq 0$. For each n choose p_n, q_n so that

$$1 \leq a_n' < 1 + \lambda |c_n|, \qquad 1 \leq d_n' < 1 + \lambda |c_n|.$$

Then $0 \leq a_n' d_n' - 1 < 2\lambda |c_n| + \lambda^2 c_n^2$ and

$$|b_n'| = \left| \frac{a_n' d_n' - 1}{c_n} \right| < 2\lambda + \lambda^2 |c_n| < 2\lambda + 2\lambda^2 |\gamma|$$

for $n > N$. The sequences a_n', b_n', and d_n' are therefore bounded and we can extract convergent subsequences. It follows there is a convergent sequence $\{V_p\}$, the elements of which are distinct since the $\{c_n\}$ were assumed distinct. Hence Γ is not discrete.

Returning to (22) we see that $\{V_n i\}$ is bounded. Denote by $V_k(i) = w_k$ a point of $\{\Gamma i\}$ of maximum imaginary part.

We shall use w_k as the center of a normal polygon N_k. Let L be the horizontal line passing through w_k. The only images of w_k lying on L are the points $w_{k,m} = w_k + m\lambda$, m = integer. The H-bisector of the H-segment $w_k w_{k,m}$ is the same as the euclidean bisector of the euclidean segment $w_k w_{k,m}$. This follows from the fact that w_k and $w_{k,m}$ are at the same height above the real axis, and is easily checked by symmetry from the formula for hyperbolic arc length. The bisectors nearest w_k bound a vertical strip S, and we are now going to show that part of S lies in N_k.

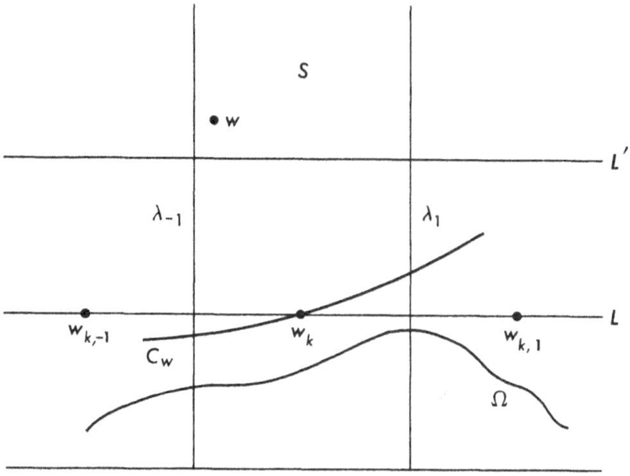

FIG. 9.

Let $\Omega = \Gamma(i) - \{w_{k,m}, m = \text{integer}\}$. The set Ω lies entirely below L. Because of the discontinuity of Γ there is a finite subset ω of Ω consisting of the points of Ω lying nearest w_k. The hyperbolic circle C_w with center $w = x + iy \in S$ passing through $w_k = x_k + iy_k$ has the equation

$$(\xi - x)^2 + (\eta - y \cosh \rho)^2 = y^2 \sinh^2 \rho,$$

where $\rho = d(w, w_k)$—see Exercise 1. The slope of the tangent to C_w at w_k is

$$\frac{x - x_k}{y_k - y \cosh \rho}$$

and this $\to 0$ as y (and therefore as ρ) tends to ∞. We see that C_w tends to the line L as $y \to \infty$. Hence for all $w \in S$ lying above a certain horizontal line L', C_w contains no point of ω and therefore no point of Ω. It follows that $w \in N_k$, for

$$d(w, w_k) < d(w, \Omega),$$

and we already knew

$$d(w, w_k) < d(w, w_{k,m}), \qquad m = \pm 1, \pm 2, \cdots;$$

hence

$$d(w, w_k) < d(w, Vw_k) \qquad \text{for all } V \in \Gamma, \qquad V \neq I.$$

That is, the part of S above L' belongs to N_k. This says that N_k contains ∞; and, in fact, ∞ is the intersection of two sides of N_k that are conjugate by the translation T. In other words, ∞ is a parabolic vertex of the transformed group Γ, hence p is a parabolic vertex of the original group. We have proved

THEOREM 2. The fixed point of a parabolic element of Γ lies on the boundary of some normal polygon.

Exercise 1. The equation of the H-circle with center $w = x + iy$ and radius ρ is

$$(\xi - x)^2 + (\eta - y \cosh \rho)^2 = y^2 \sinh^2\rho.$$

4K. In the groups with which the reader is familiar (cyclic groups, doubly periodic group, modular group), it is the case that the transformations that conjugate the sides of the fundamental region generate the group. This is true in general, as we shall now prove.

Denote by $\{V_1, V_2, \cdots\}$ the group generated by $V_1, V_2, \cdots$, that is, the set of all finite products of powers of $V_1, V_2, \cdots$.

THEOREM. Let $\{s_j, s_j' \mid j = 1, 2, \cdots\}$ be the sides of N_0 and let $T_j s_j = s_j'$ with $T_j \in \Gamma$. Then

$$\Gamma = \{T_1, T_2, \cdots\}.$$

Set $\Theta = \{T_1, T_2, \cdots\}$. We have to show that $\Gamma \subset \Theta$, the reverse inclusion being obvious.

Let V be an arbitrary element of Γ and let $V(N_0) = N^*$. Join a point in N_0 to one in N^* by a straight line L. Then L is compact and crosses a finite number of normal polygons. By making small detours if necessary, we can assume that L does not pass through a vertex. Let $N_0, N_1, \cdots, N_n = N^*$ be the polygons crossed in turn as L is described from z_0 to z_1, and let W_i be elements of Γ such that

$$W_i(N_i) = N_{i+1}, \qquad i = 0, 1, \cdots, n-1.$$

We use finite induction on n. The transformation W_0 maps a certain side s_0 of N_0 onto s_0', the side common to N_0 and N_1. Then $W_0 \in \Theta$, say $W_0 = T_{i_0}$.

Next, suppose W_1 maps s_1 (a side of N_1) onto s_1', the side common to N_1 and N_2. The sides $T_{i_0}^{-1}s_1$ and $T_{i_0}^{-1}s_1'$ are conjugate sides of N_0; let $T_{i_1} \in \Theta$ map the first on the second. Clearly $W_1 = T_{i_0}T_{i_1}T_{i_0}^{-1}$; for both transformations map s_1 on s_1' (see end of 4F). Hence $W_1 \in \Theta$.

Suppose we have shown that $W_0, W_1, \cdots, W_{k-1}$ are in Θ. Write $W = W_{k-1}W_{k-2} \cdots W_0$; note that W is in Θ and maps N_0 on N_k. Let W_k map s_k (a side of N_k) on s_k', the side common to N_k and N_{k+1}. Then $W^{-1}s_k$, $W^{-1}s_k'$ are conjugate sides in N_0; there is a $T^* \in \Theta$ such that $T^*W^{-1}s_k = W^{-1}s_k'$.

Now $W_k = WT^*W^{-1}$, for both transformations have the same effect on s_k. Hence $W_k \in \Theta$ and the induction is complete: all $W_i, i = 0, 1, \cdots, n-1$ belong to Θ. But $W_{n-1}W_{n-2} \cdots W_1 W_0$ maps N_0 on N_n and so must be identical with V. It follows that $V \in \Theta$.

It is not necessarily the case that $\{T_i\}$ is a minimal set in any sense.

5. The Hyperbolic Area of the Fundamental Region

Real discrete groups can be classified by the hyperbolic area of a normal polygon. We shall find that the area is the same for all normal polygons of a tessellation, and it does not depend on the center used in the construction of the tessellation. In fact, if we define area by the Lebesgue integral, the area is the same for all fundamental regions, where naturally we must require that the fundamental region be a Lebesgue measurable set. When the normal polygon has a finite number of sides and no free sides, its H-area is finite; in all other cases it is infinite. In the finite case the area depends on the elliptic and parabolic classes of elements in the group and on one other invariant known as the genus.

5A. We shall start out by proving the important

GAUSS-BONNET THEOREM. The hyperbolic area of a hyperbolic triangle with angles α, β, γ is finite and equal to

$$\pi - (\alpha + \beta + \gamma).$$

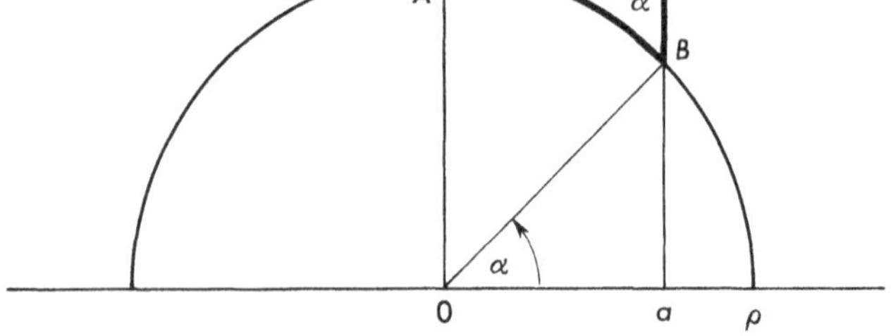

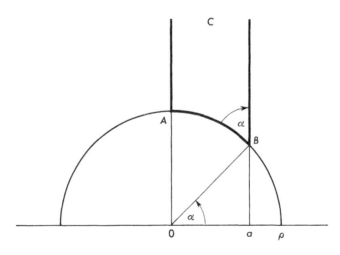

FIG. 10.

The proof is in several steps.

(1) Let the triangle be ABC with angles $0, \pi/2, \alpha$. In Figure 10 we have $0 < a \leq \rho$. Now, denoting H-area of Δ by $|\Delta|$, we have from 4B, (15),

$$|\Delta ABC| = \int_0^a dx \int_{\sqrt{\rho^2 - x^2}}^{\infty} \frac{dy}{y^2} = \int_0^a \frac{dx}{\sqrt{\rho^2 - x^2}}$$

$$= \arcsin \frac{a}{\rho} = \frac{\pi}{2} - \alpha = \pi - \left(0 + \frac{\pi}{2} + \alpha\right).$$

(2) Let ΔABC have angles $0, \alpha, \beta$, as shown in Figure 11.

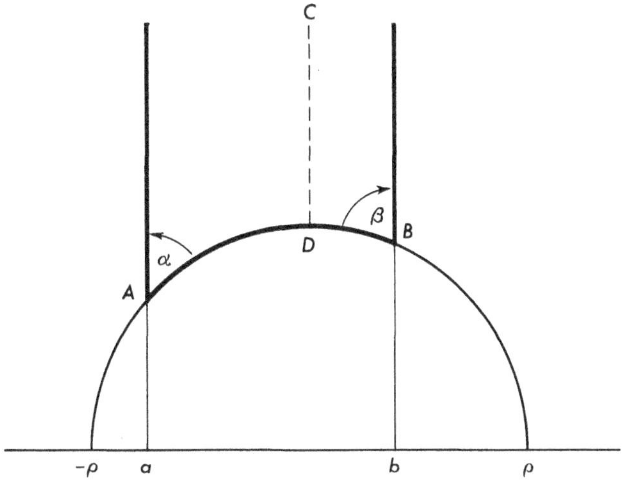

FIG. 11.

We have

$$|\Delta ADC| = \pi - \left(\alpha + \frac{\pi}{2}\right),$$

$$|\Delta DBC| = \pi - \left(\frac{\pi}{2} + \beta\right);$$

hence

$$|\Delta ABC| = 2\pi - (\pi + \alpha + \beta) = \pi - (\alpha + \beta).$$

The case in which A and B lie on the same side of D is handled similarly.

(3) If ΔABC has an infinite cusp, it is of the form (2) with $-\rho \leq a < b \leq \rho$. If it has no infinite cusp but at least one real cusp, a real linear transformation maps it with no change of hyperbolic area into a triangle with an infinite cusp.

(4) The only remaining case is that in which the triangle does not touch the real axis (see Figure 12).

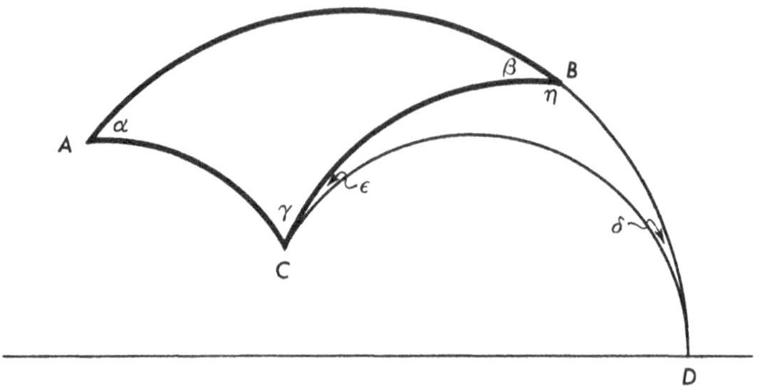

FIG. 12.

Extend AB to cut E at D. Connect C and D by an H-line. Then

$$\begin{aligned} |\Delta ABC| &= |\Delta ADC| - |\Delta BCD| \\ &= \{\pi - (\alpha + \gamma + \epsilon + \delta)\} - \{\pi - (\epsilon + \delta + \eta)\} \\ &= -(\alpha + \gamma) + (\pi - \beta). \end{aligned}$$

5B. Let us define the H-area of a set just as before except that we now insist that the integral be a Lebesgue integral.

THEOREM. Let R_1 and R_2 be fundamental regions satisfying the requirement that their boundaries (in $\bar{\mathrm{H}}$) be of Lebesgue plane measure zero. Then

$$|R_1| = |R_2|.$$

Since R_1 is open, it is Lebesgue measurable; and since Bd R_1 is of measure zero, we have $|\bar{R}_1| = |R_1|$. To prove the theorem we note the relations

$$\bar{R}_1 \supset \bigcup_{V\in\Gamma} (VR_2 \cap \bar{R}_1), \qquad \bigcup_{V\in\Gamma} (R_2 \cap V^{-1}\bar{R}_1) \supset R_2$$

of which the first is trivial and the second follows from the covering of H by images of $\bar{R}_1$. Since the terms in the right member of the first inclusion are nonoverlapping, we have

$$|R_1| = |\bar{R}_1| \geq \sum_{V\in\Gamma} |VR_2 \cap \bar{R}_1|.$$

We are here using the *complete additivity* of Lebesgue area; that is, for a countable collection of disjoint measurable sets A_n,

$$\left| \bigcup_n A_n \right| = \sum_n | A_n |.$$

By the invariance of hyperbolic area and one more application of complete additivity, we get

$$\sum_{V \in \Gamma} | VR_2 \cap \bar{R}_1 | = \sum_{V \in \Gamma} | R_2 \cap V^{-1}\bar{R}_1 | = \left| \bigcup_V \{R_2 \cap V^{-1}\bar{R}_1\} \right| \geq | R_2 |.$$

Reversing the roles of R_1 and R_2 completes the proof.

COROLLARY. Two normal polygons of Γ (in general with different centers) have equal hyperbolic area.

We have only to remark that the boundary of a normal polygon consists of a countable number of straight lines and circular arcs and is therefore of plane measure zero.

5C. THEOREM. A normal polygon has finite hyperbolic area if and only if it has a finite number of sides and no free sides.

By the preceding corollary we may consider an arbitrary normal polygon N. If N has a finite number of sides and has no free sides, its area can be computed by the Gauss-Bonnet formula. Select an interior point P in N and draw H-lines from P to the vertices of N. By the convexity of N these lines lie entirely in $\bar{N}$, which is thereby triangulated. Suppose N has $2n$ sides. Associate with each cycle a number l, which is 1 for an accidental cycle, the order of a generator of the cycle for an elliptic cycle, and ∞ (that is, $1/l = 0$) for a parabolic cycle. With this definition we see that the sum of the angles at the vertices of a cycle is in all cases equal to $2\pi l^{-1}$ (see 4H, Theorem 3). Keeping in mind that the sum of the angles about P is 2π, we now calculate,

$$| N | = 2n\pi - 2\pi - 2\pi \sum \frac{1}{l} = 2\pi \left(n - 1 - \sum \frac{1}{l} \right), \tag{23}$$

the sum being extended over all cycles. It follows that $| N |$ is finite.

For the converse, we assume $| N |$ to be finite. Then N has no free sides, since a neighborhood of an inner point of a free side clearly has infinite area.

We make an arbitrary selection of $2n$ consecutive sides $s_1, s_2, \cdots, s_{2n}$ in N. From an interior point of N draw H-lines to the end points of these sides, forming $2n$ hyperbolic triangles, whose area we shall now estimate by the Gauss-Bonnet theorem.

Let $\omega_j = \gamma_j + \beta_{j+1}$ be the angle at the vertex v_j (which may be in H or on E). Then

$$| N | \geq 2\pi n - \sum_{j=1}^{2n-1} \omega_j - \beta_1 - \gamma_{2n} - \sum_{j=1}^{2n} \alpha_j .$$

First, $\Sigma \alpha_j \leq 2\pi$. Next, $0 \leq \omega_j \leq \pi$. However, we can exclude $\omega_j = \pi$. For this case occurs only with a cycle of order 2 consisting of a single vertex and we can simply disregard this vertex, combining the two triangles on either side of it into one. Hence

$$3\pi + | N | > \beta_1 + \gamma_{2n} - \pi + 2\pi + | N | \geq \sum_{j=1}^{2n-1} (\pi - \omega_j),$$

since $\beta_1 < \omega_1 < \pi$ and similarly $\gamma_{2n} < \pi$.

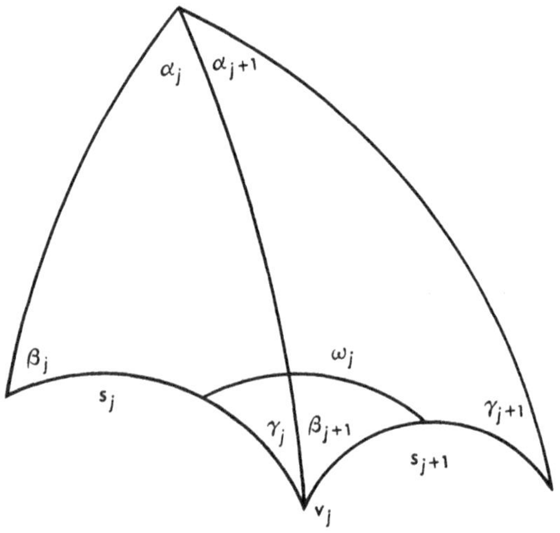

FIG. 13.

If the number of sides in N is infinite, we can extend the sequence $\{s_i\}$ indefinitely in one direction or the other; let us assume the notation is chosen so that $i \to \infty$. Then the series in the right member converges when the upper limit is replaced by infinity, since it has positive terms and its partial sums are bounded, and so $\omega_j \to \pi$. It follows that $\omega_j = 0$ for only a finite number of j, which correspond, of course, to the parabolic vertices in N. If we confine ourselves from now on to vertices in H, we can assert that

$$\tfrac{2}{3}\pi < \omega_j < \pi \qquad \text{for } j > j_0. \tag{24}$$

The vertices of N are arranged in ordinary cycles, each of which consists of a finite number of vertices. Suppose there is a cycle of order l_j ($l_j = 1$ for an accidental cycle) consisting of r_j vertices; let the vertex angles be $\omega^{(1)}, \cdots, \omega^{(r_j)}$. For $j > j_0$ we have by (24),

$$\tfrac{2}{3}\pi < \omega^{(k)} < \pi, \qquad k = 1, 2, \cdots, r_j.$$

Since

$$\omega^{(1)} + \cdots + \omega^{(r_j)} = \frac{2\pi}{l_j},$$

this gives

$$\tfrac{2}{3}\pi r_j < \frac{2\pi}{l_j} < \pi r_j,$$

or

$$2 < r_j l_j < 3,$$

a contradiction. Hence N has a finite number of sides.

COROLLARY. If $|N| < \infty$, then Γ possesses a system of generators having at most $3|N|/\pi + 6$ members.

Suppose N has $2t$ sides. The formula for the area of a normal polygon was given in (23) and is

$$|N| = 2\pi(t-1) - 2\pi \sum \frac{1}{l_j},$$

the sum being extended over all ordinary cycles. If a cycle of order l_j has r_j vertices, then $r_j l_j \geq 3$ unless $l_j = 2, r_j = 1$. This is trivial if $l_j \geq 3$ or if $l_j = 2, r_j > 1$. If $l_j = 1$ (that is, an accidental cycle), then $r_j \geq 3$. This is because the sum of the vertex angles of an accidental cycle is 2π, but no angle can be as much as π; an angle of π is associated only with a vertex of order 2. Suppose there are e cycles with $l_j = 2, r_j = 1$. Then for the remaining cycles we have

$$\sum \frac{1}{l_j} = \sum \frac{r_j}{r_j l_j} \leq \frac{1}{3} \sum r_j \leq \frac{2t - e}{3},$$

and so

$$\frac{|N|}{2\pi} \geq t - 1 - \frac{2t - e}{3} - \frac{e}{2} = \frac{t}{3} - \frac{e}{6} - 1.$$

Each of the e vertices of order 2 separate two conjugate sides that meet at the vertex. Two such vertices cannot be consecutive on the boundary of N, otherwise the side between them would have two different conjugate sides (see Figure 14). Hence $e \leq t$ and this gives

$$\frac{|N|}{2\pi} \geq \frac{t}{6} - 1.$$

Since the transformations that pair conjugate sides suffice to generate the group (4K), the proof is finished.

Exercise 1. If one normal polygon of Γ has a finite number of sides, so does every normal polygon.

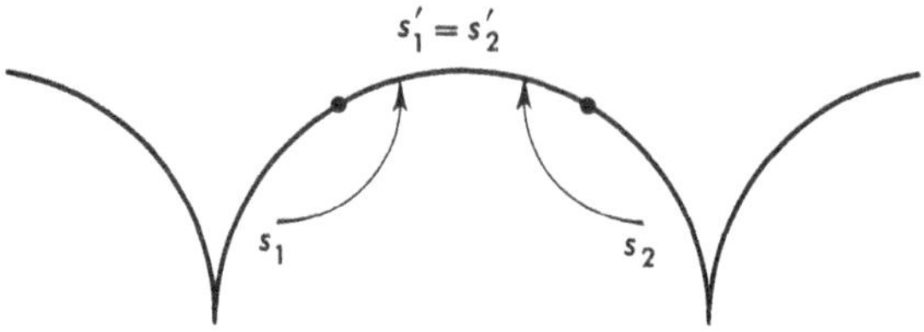

FIG. 14.

5D. THEOREM. The hyperbolic area of a fundamental region R of Γ satisfies the inequality

$$|R| \geq \frac{\pi}{21},$$

provided the boundary of R has plane measure zero.

According to our previous theorems we can restrict our arguments to a normal polygon N and can obviously assume $|N| < \infty$. Then by Theorem 5C, N has a finite number of sides and no free sides. According to (23),

$$N_1 = \frac{|N|}{2\pi} = n - 1 - \sum_{i=1}^{c} \frac{1}{l_i},$$

where N has $2n$ sides and c cycles, and N_1 is defined by this equation.

Since N has a finite number of sides, it is a polygon (4E), and we can consider it as a polygon with identifications—namely, the ones introduced by the elements of Γ. Thus N is a polygon with c vertices, n sides, and one cell. By Euler's "polyhedron formula,"

$$c - n + 1 = 2 - 2g, \tag{25}$$

where g, the *genus* of N, is a nonnegative integer. The genus is certainly a topological invariant of N, but it is not obvious that it would not change if we replaced N by a normal polygon with another center. In fact, we shall have to wait until Chapter III (Theorem 1I) for reassurance on this point. Let us assume the result here, however, and substitue (25) in the above formula for N_1, obtaining

$$N_1 = 2g - 2 + c - \sum_{i=1}^{c} \frac{1}{l_i} = 2g - 2 + \sum_{i=1}^{c} \left(1 - \frac{1}{l_i}\right).$$

In this sum we may, however, omit the accidental cycles ($l_i = 1$). Denoting by s the number of nonaccidental cycles (and relabeling if necessary) we get

$$N_1 = 2g - 2 + \sum_{i=1}^{s} \left(1 - \frac{1}{l_i}\right). \tag{26}$$

Moreover

$$N_1 > 0,$$

since N is an actual polygon. On the other hand the theorem we wish to demonstrate is equivalent to

$$N_1 \geq \frac{1}{42},$$

which we shall prove from (26) and the positivity of N_1.

The parameters in (26) are g and $\{l_1, \cdots, l_s\}$, where $g \geq 0$ and $2 \leq l_i \leq \infty$. Hence

$$\frac{1}{2} \leq 1 - \frac{1}{l_i} \leq 1.$$

For $g \geqq 2$ we have $N_1 \geq 2$. When $g = 1$, we certainly have $s > 0$, otherwise $N_1 = 0$. Hence

$$N_1 \geq 1 - \frac{1}{l_1} \geq \frac{1}{2}.$$

Suppose $g = 0$. Let $s \geq 5$; $N_1 \geq -2 + 5/2 = 1/2$. If $s = 4$, at least one term $1 - 1/l_i$ is $> 1/2$, otherwise N_1 is not > 0. If $s = 1$ or 2, we cannot fulfill $N_1 > 0$.

Therefore we are left with $s = 3$, $g = 0$ and have

$$N_1 = 1 - \left(\frac{1}{l_1} + \frac{1}{l_2} + \frac{1}{l_3}\right).$$

Let $l_1 \leq l_2 \leq l_3$. If $l_1 \geq 4$, $N_1 \geq 1 - 3 \cdot \frac{1}{4} = \frac{1}{4}$. If $l_1 \geq 3$, then since $l_2 \geq 3$ we must have $l_3 \geq 4$ in order that $N_1 > 0$. Hence $N_1 \geq 1 - (\frac{1}{3} + \frac{1}{3} + \frac{1}{4}) = \frac{1}{12}$.

Finally suppose $l_1 = 2$. Then either $l_2 = 3$ or $l_2 \geq 4$. In the first case $l_3 \geq 7$ and $N_1 \geq 1 - (\frac{1}{2} + \frac{1}{3} + \frac{1}{7}) = \frac{1}{42}$. In the second case we are forced to assume $l_3 \geq 5$. Hence $N_1 \geq 1 - (\frac{1}{2} + \frac{1}{4} + \frac{1}{5}) = \frac{1}{20}$. This concludes the proof.

We see that the lower bound is attained for $g = 0$, $l_1 = 2$, $l_2 = 3$, $l_3 = 7$. There actually is a group with these parameters. It is generated by the polygon shown in Figure 15; the generators are the transformations mating the sides connected by arrows. The proof that this group is discrete is made by quoting Poincaré's theorem, which gives sufficient conditions that a hyperbolic polygon in H give rise to a discrete group.[6]

When the group under discussion is transformed so that it acts on the unit disk U, it gives rise to the tessellation of U shown in Figure 16.

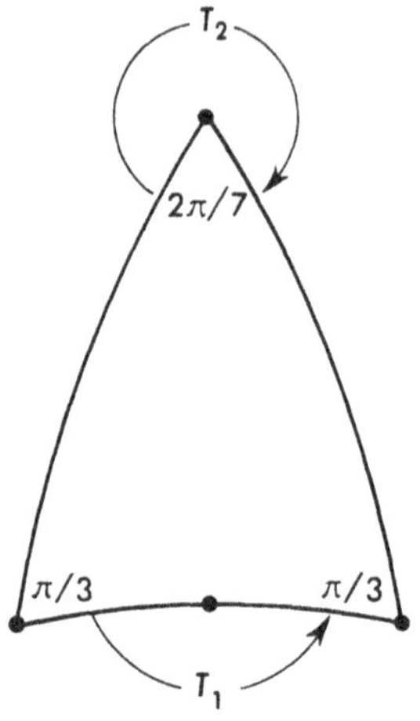

FIG. 15.

If N is not compact, at least one $l_i = \infty$. Then we can show $|N_1| \geq \frac{1}{6}$. This is the value of N_1 for the modular group. The details of the proof, which are quite similar to those in the compact case, are left to the reader.

FIG. 16. The Group (2, 3, 7). A shadad and unshaded region together constitute a fundamental region for the group. (Reprinted with the courtesy of B. G. Teubner, Stuttgart, Germany.)

Exercise 1. A group Γ is said to be *maximal* if there is no discrete group Δ in which Γ is a subgroup of finite index. For compact groups, the group with $l_1 = 2, l_2 = 3, l_3 = 7$ is maximal; for noncompact groups, the modular group is maximal.

Exercise 2. Prove that the normalizer of $\Gamma(n)$ in Ω_R is $\Gamma(1)$, the modular group (for definitions see 2J. Exercise 5, 2D).

6. Examples

In the present section we shall construct normal polygons for a number of groups.

6A. The method of bisectors, which we have exploited in the preceding sections, is not well adapted for the actual construction of normal polygons. Here we shall describe another method, associated with the name of L. R. Ford, which is based on the *isometric circle. We assume throughout that Γ contains translations.*

Let $T = (a\, b \mid c\, d) \in \Omega_R$ with $c \neq 0$. Then $|T'(z)| = |cz + d|^{-2}$. The circle

$$\mathbf{I}(T): |cz + d|^2 = 1$$

is called the isometric circle of T. Since $|T'(z)| = 1$ if and only if $z \in \mathbf{I}(T)$, the isometric circle is the locus of those points that are mapped by T without change of differential euclidean length. This explains the name. We do not define the isometric circle for T with $c = 0$.

Since T^{-1} has $c \neq 0$ when T does, $\mathbf{I}(T^{-1})$ is defined when $\mathbf{I}(T)$ is; $\mathbf{I}(T^{-1})$ is the circle $|cw - a| = 1$.

In any infinite sequence $\mathbf{I}(T_n)$ of isometric circles, the radii tend to zero. For the radius of $\mathbf{I}(T_n)$ is $1/|c_n|$, and since Γ has translations, we can apply 4J, Theorem 1.

LEMMA. T maps $\mathbf{I}(T)$ on $\mathbf{I}(T^{-1})$ and maps Ext $\mathbf{I}(T)$ on Int $\mathbf{I}(T^{-1})$.

Let z lie on or outside $\mathbf{I}(T)$; that is, $|cz + d| \geq 1$. Then

$$|cTz - a| = |cz + d|^{-1} \leq 1;$$

that is, Tz is on or inside $\mathbf{I}(T^{-1})$, and equality holds in both cases or in neither. This establishes the lemma.

Let Γ be a real discrete group. Since Γ contains translations, the stabilizer of ∞, Γ_∞, is generated by a translation

$$U^\lambda = \begin{pmatrix} 1 & \lambda \\ 0 & 1 \end{pmatrix}, \qquad \lambda > 0.$$

A fundamental region for Γ_∞ is any strip

$$R_\infty : \xi < x < \xi + \lambda.$$

Every $V \in \Gamma - \Gamma_\infty$ has an isometric circle. Define

$$R = R_\infty \cap \Big\{ \bigcup_{V \in \Gamma - \Gamma_\infty} \operatorname{Ext} \mathbf{I}(V) \Big\}. \tag{27}$$

That is, R consists of that part of R_∞ which lies exterior to every isometric circle. We shall prove that R is a fundamental region for Γ.

Suppose $z \in R$ and $T \in \Gamma$. If $T \in \Gamma_\infty$, it translates z out of R_∞, hence out of R. Otherwise z lies outside $\mathbf{I}(T)$; therefore Tz lies inside $\mathbf{I}(T^{-1})$ and so outside R. Distinct points of R are not equivalent under Γ.

We must now show that every $z \in \mathrm{H}$ is equivalent to a point of $\bar{R}$. The radii of the isometric circles are bounded above, as we have observed. Hence there is a $B > 0$ such that $z = x + iy \in \bar{R}$ if $z \in \bar{R}_\infty$ and $y \geq B$.

Let $\alpha \in \mathrm{H}$ be a boundary point of R. Then α may lie on one of the vertical sides bounding R_∞. If not, α lies on some isometric circle, for certainly α does not lie inside an isometric circle. Now α does not lie on infinitely many isometric circles, for the radii of these circles $\to 0$ and this would force α to be a point of E. A boundary point of R, then, lies on a finite number of isometric circles but inside none. It follows that a point of H not in $\bar{R}$ must lie *inside* some isometric circle.

Now let $z_0 \in \mathrm{H}$. Translate z_0 to a point z_1 in $\bar{R}_\infty$ by an element of Γ, writing

$$z_0 = x_0 + iy_0, \qquad z_1 = x_1 + iy_1$$

where $y_1 = y_0$. If z_1 is not in $\bar{R}$, it is inside some isometric circle $\mathbf{I}(V_1)$, $V_1 = (a_1\, b_1 \mid c_1\, d_1)$, and we have

$$z_2 = V_1 z_1 = x_2 + iy_2\,, \qquad y_2 = \frac{y_1}{|\, c_1 z_1 + d_1 \,|} > y_1\,.$$

Next, translate z_2 to a point z_3 in $\bar{R}_\infty$, and so on. We obtain a sequence $z_0, z_1, \cdots, z_n = x_n + iy_n, \cdots$ with

$$y_0 = y_1 < y_2 = y_3 < y_4 = \cdots.$$

If, for some n, $y_{2n+1} \geq B$, then z_{2n+1} is in $\bar{R}$ and the proof ends. Otherwise there is an infinite sequence $z_1, z_3, z_5, \cdots$, all images of z_0 under Γ and all lying in $\bar{R}_\infty$. This sequence has a point of accumulation z^* in $\bar{R}_\infty$ and definitely $\operatorname{Im} y^* > 0$, since y_{2n+1} is strictly increasing. Hence z^* is in H but its every neighborhood contains infinitely many images of z_0, a contradiction to the discontinuity of Γ.

THEOREM. The set R defined by (27) is a fundamental region for Γ.

6B. Modular Group. The modular group $\Gamma(1)$ is the group of all 2×2 matrices of determinant 1 with rational integral entries (see 2D). The stabilizer of ∞ is the cyclic group generated by

$$U = \begin{pmatrix} 1 & 1 \\ 0 & 1 \end{pmatrix}.$$

We take as R_∞ the strip $|x| < \frac{1}{2}$. The largest isometric circles are of radius 1 and have centers at the integers. Only the three centers $0, 1, -1$ intersect R_∞ and they intersect each other at $\rho = e(1/3) = -1/2 + i\sqrt{3}/2$, $-\bar{\rho} = e(1/6)$ (see Figure 17). The remaining ones have radii $\leq \frac{1}{2}$ and hence do not intrude on the region R shown. Hence

$$R : x^2 + y^2 > 1, \qquad |x| < \tfrac{1}{2}, \qquad y > 0$$

is a fundamental region for the modular group.

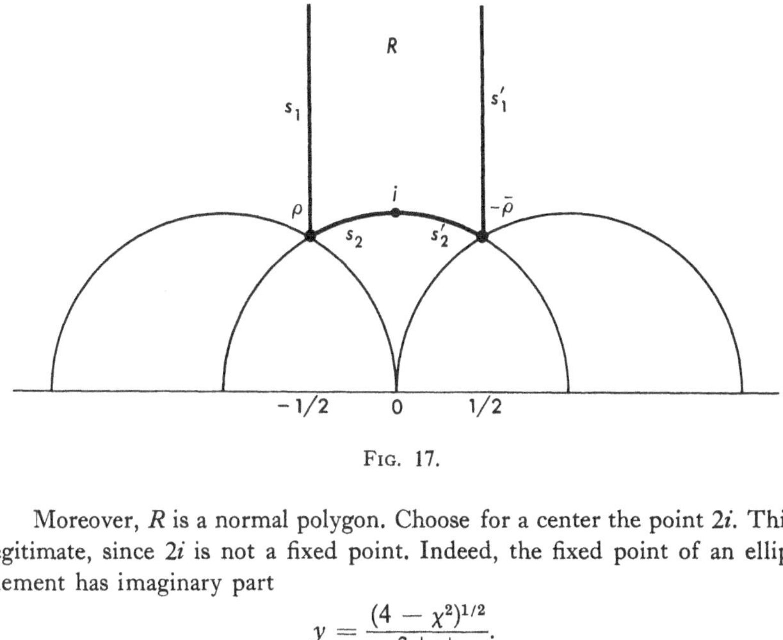

FIG. 17.

Moreover, R is a normal polygon. Choose for a center the point $2i$. This is legitimate, since $2i$ is not a fixed point. Indeed, the fixed point of an elliptic element has imaginary part

$$y = \frac{(4 - \chi^2)^{1/2}}{2\,|c|}.$$

Since $|\chi| < 2$ and an integer, we have $y \leq 2/2\,|c| \leq 1$. Now $2i + 1$, $2i - 1$ are images of $2i$; hence the lines $x = \pm\frac{1}{2}$ are H-bisectors. Also

$$T = \begin{pmatrix} 0 & -1 \\ 1 & 0 \end{pmatrix}$$

is in the group, and $T(2i) = i/2$. The H-bisector of the H-line connecting $2i$ and $i/2$—a segment of the imaginary axis—is the unit circle. Recalling the construction of the normal polygon, we see that N, the normal polygon with center $2i$, is contained in R. If $R - N$ is not empty, it is seen that $R - \bar{N}$ is not empty. Let $z \in R - \bar{N}$ and let $z' \in \bar{N}$ be equivalent to z. Then $z \neq z'$. Now $z' \in \bar{R}$. But since $z \in R$, we must have $z' \in R$, since an element of Γ never maps an interior point of R onto a boundary point. The points z and z' are distinct and equivalent, and both lie in one fundamental region. This is a contradiction and it follows that $N = R$.

A part of the tessellation of H ("modular figure") determined by R is shown in Figure 3, page 29.

The sides of R are s_1, s_1', conjugated by U, and s_2, s_2', conjugated by

$$T = \begin{pmatrix} 0 & -1 \\ 1 & 0 \end{pmatrix}.$$

Hence U and T generate the modular group. T is elliptic of order 2 and its fixed point i must be counted as a vertex, according to our convention of 4F. The cycles are $\{\rho, -\bar{\rho}\}$, $\{i\}$, and $\{\infty\}$. The sum of the angles at the vertices of the first cycle is $2\pi/3$, hence ρ is the fixed point of an elliptic element of order 3, which is in fact

$$W = \begin{pmatrix} 0 & -1 \\ 1 & 1 \end{pmatrix}.$$

Likewise $-\bar{\rho}$ is fixed by $(0\ -1 \mid 1\ -1)$. These transformations may be found by observing that TU fixes ρ, while TU^{-1} fixes $-\bar{\rho}$. The angle sum at the cycle $\{i\}$ is $\pi = 2\pi/2$, as it has to be.

There are two obvious relations in $\Gamma(1)$, namely,

$$T^2 = -I, \qquad (TU)^3 = -I.$$

It can be shown that these are defining relations for the group.[7]

6C. Subgroups of the Modular Group. The basic subgroups of the modular group are the principal congruence subgroups of level n, defined in 2D and denoted by $\Gamma(n)$. We consider here only $\Gamma(2)$, the set of modular transformations $(a\ b \mid c\ d)$ with a and d odd, b and c even.

The stabilizer of ∞ is generated by $U^2 = (1\ 2 \mid 0\ 1)$. For R_∞ take the strip $|x| < 1$. There is an isometric circle A: $|2z - 1| = 1$ and another one B: $|2z + 1| = 1$. These cut out of R_∞ the region R shown in Figure 18.

The general isometric circle is $|cz + d| = 1$, where c is even, d odd, and we may assume $c > 0$. If $c = 2$, the isometric circle does not meet R unless $d = \pm 1$—that is, unless the circle is either A or B. If $c \geq 3$, we need consider only $|d| < c$. The circles with $d = -1$, $d = -(c - 1)$ are tangent internally to A; those with $d = 1$, $d = c - 1$ are tangent internally to B; and those with the other values of d lie entirely in the complement of R. Hence R is a fundamental region for $\Gamma(2)$.

The sides s_1, s_1' are conjugated by U^2; the sides s_2, s_2' are conjugated by

$$Y = \begin{pmatrix} 1 & 0 \\ 2 & 1 \end{pmatrix}.$$

Hence $\Gamma(2) = \{U^2, Y\}$. The cycles are $\{\infty\}$, $\{-1, 1\}$, $\{0\}$—all parabolic. Examination reveals no obvious relations in $\Gamma(2)$, and in fact there are none, as can be shown (see Note 7). $\bar{\Gamma}(2)$ is a free group.† This is also true of $\bar{\Gamma}(n)$ for all $n > 1$.

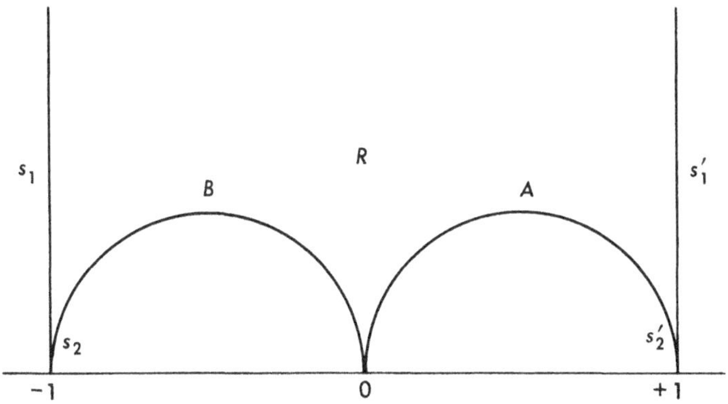

FIG. 18.

Exercise 1. Construct a fundamental region for the subgroup $\Gamma_\theta = \{U^2, T\}$ and determine its cycles. This subgroup is important in analytic number theory. [The isometric circle of T includes all isometric circles of Γ with center in $(-1, 1)$. Hence the region bounded by the lines $x = \pm 1$ and the unit circle is a fundamental region.]

Exercise 2. Assuming that every element of Γ_θ can be written uniquely as a word in U^2 and T, prove that $(a\ b \mid c\ d) \in \Gamma_\theta$ if and only if ab and cd are both even.
[Use induction on the *length* of the word.]

Exercise 3. Γ_θ is not a normal subgroup of $\Gamma(1)$.

6D. The following theorem is often handy in constructing fundamental regions for subgroups.

THEOREM. Let G be a discrete group and H a subgroup of finite index, and let

$$G = HA_1 + HA_2 + \cdots + HA_s, \qquad A_1 = I$$

† We are really talking about the *transformation* group $\bar{\Gamma}(2)$, so the relation $(-I)^2 = I$ does not count. Note that $-I \in \Gamma(2)$, but not to $\Gamma(n)$, for $n > 2$. The group $\Gamma(n)$ for $n > 2$ is free.

be a coset decomposition of H in G. If R is a fundamental region for G, then

$$R_1 = R \cup A_2(R) \cup \cdots \cup A_s(R)$$

is a fundamental region for H.

Let $z \in \mathrm{H}$. For some $V \in G$ we have $z = Vw$ with $w \in \bar{R}$. But $V = hA_i$ for an $h \in H$ and some i. Hence $z = hA_i w \in hA_i\bar{R} \subset h\bar{R}_1$, so that h^{-1} sends z into $\bar{R}_1$. Every point of H, then, is equivalent to a point of $\bar{R}_1$. Second, suppose $z, w \in R_1$ with $z = hw, h \in H$; we have $z \in A_iR, w \in A_jR$, say. Hence hA_jR overlaps A_iR in z, and since these are fundamental regions for G, we have $hA_jR = A_iR$. It follows that $hA_j = A_i$, which forces $i = j$. Thus z, w are equivalent points lying in the same fundamental region, A_iR, for G, and so are the same point. This completes the proof.

The fundamental region R_1 may not be a connected set, but it is possible to prove that the A_i can always be selected so that Int $\bar{R}_1$ is connected and is therefore a region.

Exercise 1. Construct a connected fundamental region for the group $\Gamma^0(p)$ defined by

$$\Gamma^0(p) = \left\{\begin{pmatrix} a & b \\ c & d \end{pmatrix} \in \Gamma(1) \,\middle|\, b \equiv 0 \ (\mathrm{mod}\ p)\right\},$$

where p is a prime. Show that $\Gamma^0(p)$ has exactly two parabolic cycles. (There are usually elliptic cycles, also.)
[For $p > 2$ the matrices U^j, $|j| < p/2$, and T constitute a set of coset representatives. Apply Theorem 6D.]

Exercise 2. Do the same for the group

$$\Gamma_0(p) = \left\{\begin{pmatrix} a & b \\ c & d \end{pmatrix} \in \Gamma(1) \,\middle|\, c \equiv 0 \ (\mathrm{mod}\ p)\right\}, \qquad p = \text{prime}.$$

[$\Gamma_0(p) = T\Gamma^0(p)T^{-1}$.]

Notes to Chapter 1.

1. The converse theorem, that every conformal automorphism of Z is a linear transformation, can be found in many text books. The theorem remains true if we replace Z by Ż (the plane) or by U (the open unit disk). These three regions represent, by Riemann's Mapping Theorem, all conformally equivalent classes of simply connected regions in Z.

2. Equation (5) also includes "imaginary ciicles," such as $z\bar{z} = -1$. The necessary and sufficient condition that the circle be real (that is, a locus in the complex plane) is that $B\bar{B} - AC > 0$. However, we do not have to verify this condition for the transformed circle, since the image of a real circle is real.

3. In the case of an ordinary cycle it is necessary to show that all vertices of the cycle are obtained by this procedure. We remark first that no "internal loop" can develop; that is, we never have $s_j' = s_i'$ for $j > i$. For then s_i' would be conjugate to two different sides, s_i and s_j. Second, the process cannot terminate before v_1, the initial vertex, is reached, since two sides issue from every vertex. Since an ordinary cycle has only finitely many vertices, we must eventually return to v_1. Suppose we have found the vertices $v_2, v_3, \cdots, v_t, v_1$ in that order as the result of applying our process. Let $W_j v_j = v_{j+1}, j = 1, 2, \cdots, t-1$; $W_t v_t = v_1$. The transformations $M_i = W_t W_{t-1} \cdots W_i, i = 1, 2, \cdots, t$, carry N_0 into normal polygons $N_1, N_2, \cdots, N_t$ in clockwise order about v_1 with N_i abutting $N_{i-1} (i > 1)$ and N_0 abutting N_t. See Figure 6 and the discussion on page 43. Now $M_1 = W_t \cdots W_1$ fixes v_1 and so is the identity or elliptic. In the first case $N_1 = M_1(N_0) = N_0$, in other words, the t polygons $N_0, N_t, \cdots, N_2$ make up the complete neighborhood of v_1. If there were another vertex v equivalent to v_1 and $Vv = v_1$, then VN_0 would be a normal polygon with vertex at v_1 which, however, is not one of the t polygons already mentioned. This is impossible. Second, if M_1 is elliptic of order l, the t polygons obtained in the above manner make up an elliptic sector S, and the neighborhood of v_1 is covered by the closures of S, M_1S, $\cdots$, M_1^{l-1}S, comprising tl normal polygons in all. If v were an omitted vertex, then, as before, VN_0 would not be one of the tl polygons mentioned even though it has a vertex at v_1. This shows that $\{v_1, \cdots, v_t\}$ is the complete cycle determined by v_1.

4. In this process a side may be traversed more than once. The normal polygon shown below has the same sides as the one in the text but they are arranged in a different order. Starting at p_1 and s_1 we find successive sides and vertices in the following order:

$$p_1 s_1 s_1' p_2 s_2' s_2 p_3 s_1' s_1 p_4 s_2 s_2' p_5 s_3 s_3' p_1.$$

The cycle determined by p_1 is $\{p_1, p_2, p_3, p_4, p_5\}$.

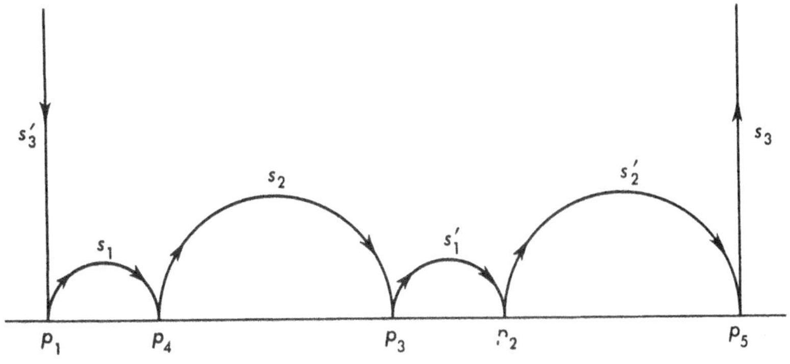

FIG. 19.

5. This result is not necessarily true if N_0 has free sides. In Figure 20, we start at p_1 and traverse the following sides and vertices in order:

$$p_1 s_1 s_1' p_2 s_2 s_2', \qquad p_3 s_1' s_1 p_4 s_2' s_2 p_5$$

This cycle does not close, since no side begins at p_5.

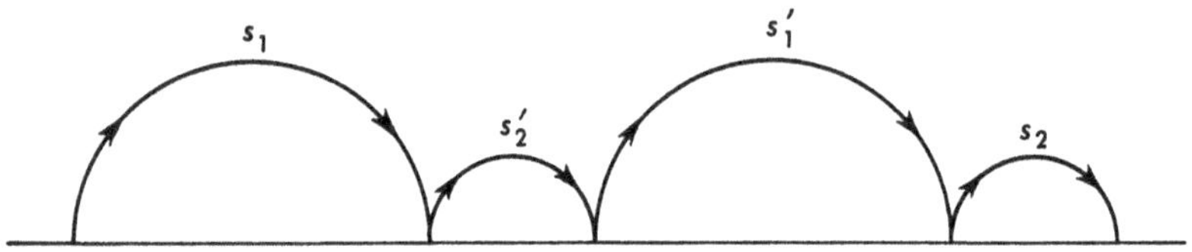

FIG. 20.

6. Poincaré's theorem is as follows. Let P be a hyperbolic polygon lying in H except possibly for cusps on E where two sides meet. Let P be bounded by a finite number of sides s_i, s_i', $i = 1, 2, \cdots, n$, and let there be given $T_i \in \Omega_R$ such that $T_i s_i = s_i'$ and such that $T_i(P)$ is disjoint from P. Let Γ be the group generated by $T_1, T_2, \cdots, T_n$. Each vertex v of P determines a cycle consisting of those vertices of P that are equivalent to v under Γ. If $v \in \mathrm{H}$ is the fixed point of an elliptic element of Γ of order l, we hypothesize that the angle sum at the vertices of the cycle determined by v shall be $2\pi/l$. If $v \in \mathrm{H}$ is not a fixed point, the angle sum shall be 2π. *With these hypotheses Poincaré concludes that Γ is discrete.*

We have seen that the above conditions are *necessary* in order that P should be a normal polygon of a discrete group. Poincaré's theorem asserts that they are *sufficient*. This theorem is of the highest importance since it enables us to construct every real discrete horocyclic group having a fundamental region with a finite number of sides. It can be extended to groups that are not horocyclic. For the proof see Fatou, pp. 125-130; Fricke-Klein, pp. 154-157; Lehner, pp. 221-227.

In the present application we have $n = 2$. There are three elliptic cycles of orders 2, 3, and 7; namely, $\{C\}$, $\{B, B'\}$, and $\{A\}$, respectively.

7. There is a theorem of Poincaré that gives a method for determining a set of defining relations for a discrete group. One draws a simple closed curve about each ordinary vertex of the fundamental region; these "loops" are made so small they do not intersect each other and each loop encloses exactly one vertex. Furthermore, it is sufficient to draw a loop around only one vertex in each cycle. Let C be a loop. It starts from a certain fundamental region R_1, crosses a finite number of other fundamental regions, say $R_2, \cdots, R_s$, and returns to R_1. If T_i is a group element mapping R_i on R_{i+1}, then $T_s \cdots T_2T_1$ maps R_1 on itself and so

$$T_s \cdots T_2T_1 = \pm I$$

is a relation in the group. Poincaré proved that the relations obtained in this way constitute a set of defining relations for the group and that it is a minimal set in a certain sense. For the proof see Fatou, pp. 188-190; Lehner, pp. 230-234; Macbeath, pp. 25-34.

[II]

Automorphic Functions and Automorphic Forms

In this chapter every real discrete group considered will have translations other than the identity. From Section 1F on, every group will be horocyclic. In Sections 2, 3, 4, all groups will have fundamental regions with a finite number of sides.

1. Existence

1A. Given a real discrete group Γ, we consider the functional equation

$$f(Vz) = f(z) \qquad \text{for all } V \in \Gamma \text{ and all } z \in \mathrm{H}. \tag{1}$$

For the time being we shall define an *automorphic function on* Γ to be any meromorphic function that is a solution of (1). Later we shall add another requirement.

Instead of discussing (1) directly, Poincaré treated a simpler problem whose solution leads immediately to a solution of (1).

Differentiating (1) we get†

$$f'(Vz) = \frac{d}{dVz} f(z) = \left(\frac{dz}{dVz}\right) f'(z) = (cz + d)^2 f'(z), \qquad V = \begin{pmatrix} \cdot & \cdot \\ c & d \end{pmatrix}.$$

In general, a meromorphic function $g(z)$ satisfying

$$g(Vz) = (cz + d)^{2m} g(z) \qquad \text{for } V \in \Gamma \text{ and } z \in \mathrm{H} \tag{2}$$

is called an *automorphic form of dimension* $-2m$. If g_1, g_2 are two automorphic forms of the same dimension and $g_2(z) \not\equiv 0$, then

$$f(z) = \frac{g_1(z)}{g_2(z)}$$

is an automorphic function.

Exercise 1. A nonconstant automorphic function has the real axis as a natural boundary if Γ is horocyclic.

---

† $f'(Vz)$ means

$$\frac{d}{dVz} f(Vz).$$

Exercise 2. The same assertion is valid for a nonconstant automorphic form.

Exercise 3. A constant automorphic form of nonzero dimension is identically zero.

Poincaré constructed an automorphic form by utilizing the mapping $\Gamma L \leftrightarrow \Gamma$ for $L \in \Gamma$. Consider

$$g(z) = \sum_{V \in \Gamma} \left(\frac{dVz}{dz}\right)^m,$$

where m is a positive integer. For the moment we set aside the question of convergence of the series and its rearrangement. Formally we have, for $L \in \Gamma$,

$$g(Lz) = \sum_{V} \left(\frac{dVLz}{dLz}\right)^m = \left(\frac{dz}{dLz}\right)^m \sum_{V} \left(\frac{dVLz}{dz}\right)^m$$

$$= (\gamma z + \delta)^{2m} g(z), \qquad L = \begin{pmatrix} \cdot & \cdot \\ \gamma & \delta \end{pmatrix}$$

since $\Gamma L = \Gamma$. Thus $g(z)$ satisfies (2).

However, the question of convergence is not trivial. Let Γ_∞, the subgroup of translations in Γ, be generated by $T = (1\ \lambda \mid 0\ 1)$. The above series can be written

$$\sum_{V \in \Gamma} (cz + d)^{-2m}, \qquad V = \begin{pmatrix} \cdot & \cdot \\ c & d \end{pmatrix}.$$

The summand $(cz + d)^{-2m}$ occurs infinitely often, since Γ contains with V also $T^m V = (\cdot\ \cdot \mid c\ d)$ for each integer m, a situation not calculated to produce convergence.

We therefore consider a sum extended over a single representative from each class $\{T^m V\}$. If we write the coset decomposition

$$\Gamma = \Gamma_\infty + \Gamma_\infty V_1 + \Gamma_\infty V_2 + \cdots,$$

the sum will be over the $\{V_i\}$, or, as we shall write, over a system of representatives of[†] Γ_∞ / Γ. Since we shall need a whole family of automorphic forms, we throw in a general factor and consider

$$g(z) = \sum_{V_i \in (\Gamma_\infty \backslash \Gamma)} H(V_i z) \cdot (c_i z + d_i)^{-2m}, \qquad V_i = \begin{pmatrix} \cdot & \cdot \\ c_i & d_i \end{pmatrix},$$

where $H(z)$ is a regular function in H. This definition makes sense only if the

[†] If H is a subgroup of G, we write (G/H) for a system of representatives of the cosets $\{xH\}$. Likewise, we write $(H \backslash G)$ for a system of representatives of the cosets $\{Hy\}$.

sum of the series is independent of the choice of representatives. Suppose instead of V_i we had used $T^{m_i}V_i$; we would have

$$g_1(z) = \sum_{V_i} H(T^{m_i}V_i z)(c_i z + d_i)^{-2m},$$

and $g = g_1$ if

$$H(T^{m_i}V_i z) = H(V_i z + m_i\lambda) = H(V_i z).$$

This condition will certainly be satisfied if H is periodic with period λ, for example,[†]

$$H(z) = e\left(\frac{\nu z}{\lambda}\right), \qquad \nu = \text{integer}.$$

Then the series we wish to consider is

$$G_{-2m}(z, \nu) = \sum_{M\in(\Gamma_\infty\backslash\Gamma)} \frac{e(\nu Mz/\lambda)}{(cz + d)^{2m}}, \qquad M = \begin{pmatrix} a & b \\ c & d \end{pmatrix}. \tag{3}$$

This is called a Poincaré series of dimension $-2m$ and parameter ν. For our application we shall need to establish its absolute convergence, for this will permit rearrangement of its terms.

1B. We shall now proceed to a systematic study of the Poincaré series. Let $\Gamma_\infty = \{U^\lambda\}$, $\lambda > 0$, where we define the symbolic power

$$U^\lambda = \begin{pmatrix} 1 & \lambda \\ 0 & 1 \end{pmatrix}, \qquad \lambda \text{ real}.$$

It is convenient to assume that $-I \in \Gamma$; if it does not, we replace Γ by the group generated by $-I$ and Γ. Note, however, that Γ_∞ does not contain $-I$.

THEOREM. The series

$$\sum_{M_i\in(\mathbf{D})} | c_i\tau + d_i |^{-r}, \qquad M_i = \begin{pmatrix} a_i & b_i \\ c_i & d_i \end{pmatrix}, \qquad \mathbf{D} = \Gamma_\infty\backslash\Gamma \tag{4}$$

converges when $r > 2$ for each $\tau = x + iy \in \mathrm{H}$, and converges uniformly on each closed region

$$E_\alpha\colon | x | \leq \alpha^{-1}, \qquad y > \alpha > 0.$$

In particular, the convergence is uniform on compact subsets of H. (Here r is not necessarily an integer.)

[†] We write, throughout,

$$e(u) = e^{2\pi i u}.$$

D is a complete system of matrices of Γ with different lower row. Since $-I \notin \Gamma_\infty$, M and $-M$ appear in different cosets of **D**. We write

$$\sum_{M_i \in (\mathbf{D})} | c_i\tau + d_i |^{-r} = 2 \sum_{M_i \in (\mathbf{D}')} | c_i\tau + d_i |^{-r},$$

where $(\mathbf{D}')$ consists of those M_i in $(\mathbf{D})$ for which $c_i > 0$, or $c_i = 0$ and $a_i > 0$. Different M_i in $(\mathbf{D}')$ correspond to different linear transformations $\bar{M}_i\tau$ and to different normal polygons M_iN_0.

Let N_0 be an unbounded normal polygon bounded laterally by portions of the lines $x = \pm\lambda/2$. Such a polygon can be obtained by using as a center a point on the imaginary axis of sufficiently large imaginary part. Note also that *any* normal polygon lies within some vertical strip of width λ, since, whatever its center may be, the vertical lines on either side distant $\lambda/2$ are perpendicular bisectors used in its construction. Hence any normal polygon can be translated by a power of U^λ to a polygon lying in the strip S: $-\lambda < x < \lambda$.

Since the sum of the series (4) is independent of the choice of $(\mathbf{D}')$, as we have seen, select $(\mathbf{D}')$ so that M_iN_0 lies in S for all $M_i \in \mathbf{D}$. The transformation M_0 is the unique M_i with $c_i = 0$ and necessarily $M_0 = I$.

The polygon N_0 is bounded below by arcs of circles and possibly by segments of the real axis. These arcs all lie below a certain horizontal line, for ∞ is a parabolic vertex and a certain region $| x | < \lambda/2, y > Y$ must lie in N_0. Let K_0 be a disk in N_0 lying above this line. It is clear that $M_i(K_0) = K_i$ is a disk lying nearer to the real axis than K_0. Thus all K_i lie below a certain horizontal line $y = \beta$. Moreover, the K_i lie in different normal polygons $N_i = M_i(N_0)$ and so are disjoint. The rectangle $| x | \leq \lambda$, $0 < y < \beta$ contains all K_i.

Define

$$J = \int_0^\beta \int_{-\lambda}^{\lambda} y^{r/2} \frac{dx\, dy}{y^2};$$

the integral converges when $2 - r/2 < 1$, that is, when $r > 2$. For each positive integer N,

$$J > \sum_{i=0}^{N} \iint_{K_i} y^{r/2} \frac{dx\, dy}{y^2}.$$

In each integral of this sum we make the change of variable† $\tau = M_i\omega$, $\omega = u + iv$. Using $y = v\, | c_i\omega + d_i |^{-2}$ and the invariance of the area element, we get

$$J > \sum_{i=0}^{N} \iint_{K_0} v^{r/2} | c_i\omega + d_i |^{-r} \frac{du\, dv}{v^2}.$$

† The change of variable is legitimate since M_i is a one-to-one mapping of K_0 on K_i whose real and imaginary parts are continuously differentiable.

If ω_0 is the center of K_0, it is easy to check that

$$\left| \frac{c_i\omega + d_i}{c_i\omega_0 + d_i} \right| \leq C_1, \qquad \omega \in K_0$$

where $C_1, C_2, \cdots$ denote positive constants depending at most on K_0, α, and r. This inequality is trivial for $c_i = 0$, and for $c_i \neq 0$ we have, with ρ the radius of K_0,

$$\left| \frac{c_i\omega + d_i}{c_i\omega_0 + d_i} \right| = \left| \frac{\omega + d_i/c_i}{\omega_0 + d_i/c_i} \right| \leq \frac{|\omega - \omega_0| + |\omega_0 + d_i/c_i|}{|\omega_0 + d_i/c_i|}$$

$$\leq 1 + \frac{\rho}{\operatorname{Im} \omega_0} = C_1,$$

since $-d_i/c_i$ is real and so $|\omega_0 + d_i/c_i| \geq \operatorname{Im} \omega_0$.

Hence for $r > 2$,

$$\infty > J > \sum_{i=0}^{N} |c_i\omega_0 + d_i|^{-r} C_1^{-r} \iint_{K_0} v^{r/2-2}\, dv = C_2 \sum_{0}^{N} |c_i\omega_0 + d_i|^{-r}.$$

It follows that the series (4) converges at $\tau = \omega_0$.

LEMMA. If $\tau = x + iy$, $\omega_0 = x_0 + iy_0$ lie in E_α, then

$$|c\tau + d| \geq A\, |c\omega_0 + d|$$

for all real c, d, where A depends only on α and ω_0.

Let $A_1, A_2, \cdots$ denote constants depending at most on α and ω_0. Since the ratio of $|c\omega_0 + d|$ to $|ci + d|$ is obviously bounded above by A_1, we may replace the inequality by

$$|c\tau + d|^2 \geq A_2(c^2 + d^2).$$

Disregarding the trivial case $c = 0$, we first assume $|d/c| \leq 2/\alpha$ and have

$$c^2 + d^2 \leq c^2(1 + 4\alpha^{-2}) = \frac{1 + 4\alpha^{-2}}{\alpha^2} \cdot c^2\alpha^2 = A_3c^2\alpha^2 \leq A_3c^2 \left| \tau + \frac{d}{c} \right|^2.$$

On the other hand, if $|d/c| > 2/\alpha$, we get $|x + d/c| \geq |d/c| - 1/\alpha > |d/2c|$; hence

$$|c\tau + d|^2 = c^2\left(\left(x + \frac{d}{c}\right)^2 + y^2\right) \geq \frac{d^2}{4} + c^2\alpha^2 \geq A_4(c^2 + d^2),$$

where $A_4 = \min(\frac{1}{4}, \alpha^2)$. This establishes the lemma.

Using the lemma we get

$$\sum_{i=0}^{\infty} |c_i\tau + d_i|^{-r} \leq C_3 \sum_{i=0}^{\infty} |c_i\omega_0 + d_i|^{-r}$$

uniformly for all $\tau \in E_\alpha$. This concludes the proof of the theorem.

Exercise 1. The series (4) converges at every ordinary point of Γ and converges uniformly on every compact set whose distance from the limit set of Γ is positive.

1C. We are now in a position to discuss the Poincaré series (3).

THEOREM. The Poincaré series $G_{-r}(\tau, \nu)$ for $r > 2$ converges absolutely uniformly† on regions E_α, $\alpha > 0$. When r is an integer and $\nu \geqslant 0$, its sum is a meromorphic (actually regular) function in H and satisfies the functional equation

$$G_{-r}(M\tau, \nu) = (c\tau + d)^r G_{-r}(\tau, \nu), \qquad M = \begin{pmatrix} \cdot & \cdot \\ c & d \end{pmatrix} \in \Gamma. \tag{5}$$

When $\nu \leq 0$ and r is even, $G_{-r}(\tau, \nu)$ is not a constant.

We have, for $\tau = x + iy \in E_\alpha$ and $M_i = (\cdot\ \cdot \mid c_i\ d_i)$, $c_i \neq 0$,

$$\left| e\left(\frac{\nu M_i\tau}{\lambda}\right) \right| \leq \exp\left\{\frac{2\pi|\nu|y/\lambda}{(c_i x + d_i)^2 + c_i^2 y^2}\right\} \leq \exp\left\{\frac{2\pi|\nu|}{c_i^2 y\lambda}\right\} \leq \exp\left\{\frac{2\pi|\nu|}{\tilde{c}^2\alpha\lambda}\right\},$$

where $\tilde{c} > 0$ is a lower bound for $|c_i| \neq 0$ (I, 4J, Theorem 1). If $c_i = 0$, we may choose $M_i = \pm I$, and for $\nu \geq 0$ we have

$$\left| e\left(\frac{\nu M_i\tau}{\lambda}\right) \right| \leq \exp\left(\frac{-2\pi\nu y}{\lambda}\right) \leq 1.$$

Thus for $\nu \geq 0$ all terms of the series are majorized by $|c_i\tau + d_i|^{-r}$. For $\nu < 0$ this is true of all terms except the two with $M_i = \pm I$; these terms may be disregarded in considering the convergence of the series. By Theorem 1B the series converges as asserted.

When r is an integer and $\nu \geq 0$, each term of the series is regular in H, and the uniform convergence on compact subsets of H shows that $G_{-r}(\tau, \nu)$ is regular in H.

We now prove the functional equation. Let

$$L = \begin{pmatrix} \alpha & \beta \\ \gamma & \delta \end{pmatrix} \in \Gamma, \qquad M_i L = \begin{pmatrix} \cdot & \cdot \\ c' & d' \end{pmatrix}.$$

We have

$$(c'\tau + d')^r = (c_i L\tau + d_i)^r (\gamma\tau + \delta)^r.$$

† That is, the sum of the absolute values converges uniformly.

Hence

$$G_{-r}(L\tau, \nu) = \sum_{M_i \in (\mathbf{D})} \frac{e(\nu M_i L\tau/\lambda)}{(c_i L\tau + d_i)^r}$$

$$= (\gamma\tau + \delta)^r \sum_{M_i \in (\mathbf{D})} \frac{e(\nu M_i L\tau/\lambda)}{(c'\tau + d')^r}.$$

Now $M_i L$ runs over $(\Gamma_\infty \backslash \Gamma)$ if M_i does. For suppose $M_1 L$ and $M_2 L$ have the same second row; then $(M_1 L)(M_2 L)^{-1} = (\cdot \ \cdot \mid 0 \ \cdot) \in \Gamma_\infty$, hence $M_1 L = U^{t\lambda} M_2 L$, or $M_1 = U^{t\lambda} M_2$. That is, M_1 and M_2 are in the same coset of $\Gamma_\infty \backslash \Gamma$, a contradiction. Absolute convergence permits us to rearrange the series, and it follows that the right member of the above equation is

$$(\gamma\tau + \delta)^r G_{-r}(\tau, \nu),$$

which concludes the proof that G_{-r} satisfies the functional equation.

The final statement of the theorem is proved as follows. Suppose $\nu < 0$; we have

$$G_{-r}(\tau, \nu) = \sum_1 + \sum_2,$$

where Σ_1 contains the terms with $c_i = 0$, that is, $M_i = \pm I$, and Σ_2 consists of the terms with $c_i \neq 0$. Now Σ_2 converges uniformly on compact subsets of H and its sum is regular there. But since r is even,

$$\sum_1 = 2e^{2\pi i\nu\tau/\lambda} = 2e^{-2\pi\nu y/\lambda} \cdot e^{2\pi i\nu x/\lambda}, \qquad \tau = x + iy$$

and $|\Sigma_1| \to +\infty$ as $\tau \to i\infty$ (that is, $|x| \leq \lambda/2$, x fixed, $y \to +\infty$). Hence

$$|G_{-r}(\tau, \nu)| \to \infty \qquad \text{as } \tau \to i\infty, \qquad \nu < 0$$

and $G_{-r}(\tau, \nu)$ cannot be constant. Finally, consider $\nu = 0$. By uniform convergence

$$\lim_{\tau \to i\infty} G_{-r}(\tau, 0) = \sum_{M \in (\mathbf{D})} \lim_{\tau \to i\infty} \frac{1}{|c\tau + d|^r} = 0 + \sum_{M_i = \pm I} \lim_{\tau \to i\infty} 1 = 2.$$

If $G_{-r}(\tau, 0) = C$, a constant, (5) shows that $C = 0$. But $G_{-r}(i\infty, 0) \neq 0$. Hence, $G_{-r}(\tau, 0)$ is nonconstant and the theorem is proved.

The theorem asserts that $G_{-r}(\tau, \nu)$ is an automorphic form of dimension $-r$. Note that $G_{-r}(\tau, \nu) \equiv 0$ if r is odd, for Γ contains both M_i and $-M_i$.

1D. THEOREM. If Γ is a real discrete group possessing translations, there exist nonconstant automorphic functions on Γ.

The function

$$f(\tau) = \frac{G_{-4}(\tau, -1)}{G_{-4}(\tau, 0)}$$

is invariant on Γ, meromorphic in H, and $|f(\tau)| \to \infty$ as $\tau \to i\infty$.

1E. We now take up the question of how the Poincaré series behaves at points of E. We assume r even. As before, write

$$G_{-r}(\tau, \nu) = \sum_1 + \sum_2,$$

where Σ_1 contains the terms M_i for which $c_i = 0$ (that is, $\pm I$). As we have seen, $\Sigma_2 \to 0$ with $\tau \to i\infty$. Since $\Sigma_1 = 2e(\nu\tau/\lambda)$, we have

$$\lim_{\tau\to i\infty} G_{-r}(\tau, \nu) = \begin{cases} 0, & \nu > 0 \\ 2, & \nu = 0 \\ \infty, & \nu < 0 \end{cases}.$$

Therefore $G_{-r}(\tau, \nu)$ tends to a definite limit (finite or infinite) as $\tau = x + iy \to i\infty$, uniformly in x.

How does G_{-r} behave when τ approaches a finite parabolic cusp? In order to answer this question, we shall introduce the *S-transform of an automorphic form.*

Denote by $\{\Gamma, -r\}$ the set of all automorphic forms on Γ of dimension $-r$. Obviously $\{\Gamma, -r\}$ is a vector space over the complex numbers. The functions $G_{-r}(\tau, \nu) \in \{\Gamma, -r\}$.

We define the S-transform of $F(\tau)$ by

$$F_S(\tau) = F \mid S = \left(\frac{dS^{-1}\tau}{d\tau}\right)^{r/2} F(S^{-1}\tau) = (g\tau + h)^{-r} F(S^{-1}\tau),$$

where $S^{-1} = (\cdot\ \cdot \mid g\ h) \in \Omega_R$.[†] The transformed group $S\Gamma S^{-1}$ is a real discrete group. The set of parabolic vertices of $S\Gamma S^{-1}$ is SP, where P is the set of parabolic vertices of Γ. Thus $S\Gamma S^{-1}$ may not have translations.

LEMMA 1. If $F \in \{\Gamma, -r\}$, then

$$F \mid S \in \{S\Gamma S^{-1}, -r\}.$$

The functional equation for an automorphic form (2) can be written as

$$F(V\tau)(dV\tau)^{r/2} = F(\tau)(d\tau)^{r/2}, \qquad V \in \Gamma.$$

[†] In this notation the transformation equation (2) can be written $g \mid V^{-1} = g$.

Hence

$$F_S(SLS^{-1}\tau)(dSLS^{-1}\tau)^{r/2} = F(LS^{-1}\tau)\left(\frac{dS^{-1}SLS^{-1}\tau}{dSLS^{-1}\tau}\right)^{r/2}(dSLS^{-1}\tau)^{r/2}$$

$$= F(L[S^{-1}\tau])(dL[S^{-1}\tau])^{r/2} = F(S^{-1}\tau)(dS^{-1}\tau)^{r/2}$$

$$= F_S(\tau)(d\tau)^{r/2},$$

as required. It is clear that $F \mid S$ is meromorphic in H.

We now define: F is regular at $S^{-1}\tau$ if F_S is regular at τ; F has a pole of order ν at $S^{-1}\tau$ if F_S has a pole of order ν at τ; etcetera. The reader should verify that the definition is independent of the choice of S; that is, if $T^{-1}\tau = S^{-1}\tau$, then F_S and F_T have the same behavior at τ.

Suppose p is a finite parabolic cusp of Γ. Define

$$A_p = A = \begin{pmatrix} 0 & -1 \\ 1 & -p \end{pmatrix}.$$

Thus $A_p(p) = \infty$. Let P be a generator of Γ_p. Then $P' = APA^{-1}$ fixes ∞ and is parabolic, since P is. Hence P' is a translation. Since $(APA^{-1})^{-1} = AP^{-1}A^{-1}$, we can replace P by its inverse if necessary so as to have

$$P' = U^{\lambda'} = \begin{pmatrix} 1 & \lambda' \\ 0 & 1 \end{pmatrix}, \qquad \lambda' > 0.$$

P' belongs to the group $\Gamma' = A\Gamma A^{-1}$ and in fact P' generates Γ'_∞. Note that Γ' is real discrete.

Let us first dispose of the case in which p is Γ-equivalent to ∞; $Vp = \infty$. We must consider

$$G_{-r}(\tau, \nu)\mid V = (-c\tau + a)^{-r} G_{-r}(V^{-1}\tau, \nu)$$

at $\tau = i\infty$. But by the transformation equation of an automorphic form we have

$$G_{-r}(\tau, \nu)\mid V = G_{-r}(\tau, \nu),$$

so G_{-r} has the same behavior at all vertices equivalent to ∞.

Hence we now assume p is not equivalent to ∞ and is therefore finite. We must consider

$$G_{-r}(\tau, \nu)\mid A = (-\tau)^{-r} G_{-r}(A^{-1}\tau, \nu) \tag{6}$$

at $\tau = i\infty$. The right member equals

$$(-\tau)^{-r} \sum_{M \in (\Gamma_\infty \backslash \Gamma)} \frac{e(\nu M A^{-1}\tau/\lambda)}{(cA^{-1}\tau + d)^r}, \qquad M = \begin{pmatrix} \cdot & \cdot \\ c & d \end{pmatrix}.$$

Since

$$(cA^{-1}\tau + d)(-\tau) = c'\tau + d',$$

where

$$MA^{-1} = M_1 = \begin{pmatrix} \cdot & \cdot \\ c' & d' \end{pmatrix},$$

we get

$$G_{-r}(\tau, \nu) \mid A = \sum_{M_1 \in (\mathbf{S})} \frac{e(\nu M_1 \tau/\lambda)}{(c'\tau + d')^r}. \tag{7}$$

Here M_1 runs over $(\mathbf{S}) = (\Gamma_\infty \backslash \Gamma A^{-1})$, which, as is easily verified, is a complete system of matrices in ΓA^{-1} with different lower row.

The matrices in $(\mathbf{S})$ never have $c' = 0$. Indeed, this would imply that $M_1(\infty) = MA^{-1}(\infty) = \infty$, that is, $Mp = \infty$, whereas we assumed p not equivalent to ∞. For the purpose of estimating the numerators of the series in (7), we need a lemma.

LEMMA 2. *The matrices* $(\alpha\, \beta \mid \gamma\, \delta)$ *of* ΓA^{-1} *have the property that the nonzero* γ*'s are bounded below in absolute value by a positive constant.*

The proof is substantially the same as in the previous case, when we considered Γ instead of ΓA^{-1} (I, 4J, Theorem 1). Let $\gamma_n \to 0$, γ_n distinct, where each γ_n occurs in $X_n = (\alpha_n\, \beta_n \mid \gamma_n\, \delta_n) \in \Gamma A^{-1}$. We may assume $\gamma_n \neq 0$ and in fact $\gamma_n > 0$ (otherwise replace X_n by $-X_n$). Write $X_n = M_n A^{-1}$, $M_n \in \Gamma$. Then

$$Y_n = U^{s_n\lambda} M_n P^{t_n} A^{-1} = U^{s_n\lambda} X_n P'^{t_n}$$

is an element of ΓA^{-1} for arbitrary integers s_n, t_n, where $P' = APA^{-1} = U^{\lambda'} \in \Gamma'_\infty$. We calculate

$$Y_n = \begin{pmatrix} 1 & s_n\lambda \\ 0 & 1 \end{pmatrix} \begin{pmatrix} \alpha_n & \beta_n \\ \gamma_n & \delta_n \end{pmatrix} \begin{pmatrix} 1 & t_n\lambda' \\ 0 & 1 \end{pmatrix} = \begin{pmatrix} \alpha_n' & \beta_n' \\ \gamma_n & \delta_n' \end{pmatrix}$$

with

$$\alpha_n' = \alpha_n + s_n\gamma_n\lambda, \qquad \delta_n' = \delta_n + t_n\gamma_n\lambda'.$$

As in the earlier proof we can select integers s_n, t_n so that on a subsequence (m) we have $Y_m \to Y$, Y being a unimodular real matrix. If we set $T_m = Y_{m+1}^{-1} Y_m$, we then find

$$T_m = AW_m A^{-1},$$

with $W_m \in \Gamma$. That is, $T_m \in A\Gamma A^{-1}$, a real discrete group. But $T_m \to I$. If $T_m = I$ for $m > m_0$, we would have $Y_m = Y_{m+1} (m > m_0)$, and this contradicts the fact that the $\{\gamma_m\}$ are all different. This concludes the proof of the lemma.

We now estimate

$$\left| e\left(\frac{\nu M_1\tau}{\lambda}\right)\right| \leq \exp\left\{\frac{2\pi|\nu|}{\lambda}\frac{y}{(c'x+d')^2+c'^2y^2}\right\} \leq \exp\left\{\frac{2\pi|\nu|}{\tilde{c}'^2 y\lambda}\right\},$$

where $\tilde{c}'$ is a positive lower bound for the moduli of the nonzero c' in the matrices of ΓA^{-1}. For $\tau \in E_\alpha$, then, the series $G_{-r}(\tau, \nu) \mid A$ (see (7)) is majorized by

$$G_{-r}(\tau, 0) \mid A = \sum_{M_1 \in (S)} (c'\tau + d')^{-r}, \qquad M_1 = \begin{pmatrix} \cdot & \cdot \\ c' & d' \end{pmatrix}.$$

From (6) we deduce that the last series converges absolutely for each $\tau \in \mathrm{H}$, for $A^{-1}\tau \in \mathrm{H}$. Its *uniform* convergence in regions E_α is shown by using the inequality

$$| c'\tau + d' | \geq A \mid c'i + d' |,$$

where A depends only on α (see Lemma 1B). Thus $G_{-r}(\tau, \nu) \mid A$ converges absolutely uniformly in E_α.

From this and the fact that $c' \neq 0$ in ΓA^{-1}, it follows that

$$G_{-r}(\tau, \nu) | A \to 0 \qquad \text{as} \qquad \tau \to i\infty. \tag{8}$$

According to our convention, (8) means $G_{-r}(\tau, \nu)$ has a zero at $\tau = p$. We have proved:

THEOREM. The Poincaré series $G_{-r}(\tau, \nu)$ has a zero at every parabolic cusp not equivalent to ∞. At $i\infty$ and at all parabolic vertices equivalent to ∞, G_{-r} has a zero if $\nu > 0$, $| G_{-r} | \to \infty$ if $\nu < 0$, and $G_{-r}(i\infty, 0) = 2$.

In (6) replace τ by $A\tau$:

$$G_{-r}(A\tau, \nu) | A = (\tau - p)^r G_{-r}(\tau, \nu). \tag{6a}$$

It appears from this equation that $(\tau - p)^r G_{-r}(\tau, \nu) \to 0$ as $\tau \to p$ vertically.

1F. We shall take the Poincaré series as our model and so we now make our final definition. *We shall, however, restrict ourselves from now on to groups of the first kind, that is,* $\mathrm{L}(\Gamma) = \mathrm{E}$.

DEFINITION. We say $F(\tau) \in \{\Gamma, -r\}$, $r =$ integer, if and only if

(i) F is meromorphic in H

(ii) $F(V\tau) = (c\tau + d)^r F(\tau)$, $V \in \Gamma$, $\tau \in \mathrm{H}$

(iii) $(\tau - p)^r F(\tau)$ tends to a definite value (which may be infinite) as $\tau \to p$, a finite parabolic vertex, and $F(\tau)$ tends to a limit as $\tau \to i\infty$. In both cases the approach is to be from within a normal polygon.†

An automorphic function is a member of $\{\Gamma, 0\}$.

Since our proof of the existence of automorphic functions was made by considering quotients of Poincaré series, it is clear that we now have:

THEOREM. If Γ is a real discrete horocyclic group possessing translations, there exist nonconstant automorphic forms on Γ of all even dimensions; in particular, there exist nonconstant automorphic functions.

To get forms of positive dimension we note that the reciprocal of an automorphic form of dimension $-r$ is one of dimension r.

It is to be observed that the only real points at which the behavior of an automorphic form is specified are parabolic vertices. This will be motivated in the next chapter.

1G. According to the notation we have just introduced the set of automorphic functions on Γ is denoted by $\{\Gamma, 0\}$. This set is not only a vector space but actually a *field*, which we denote also by $\mathbf{K}(\Gamma)$. This is no longer true, of course, for $\{\Gamma, -r\}$ when $r \neq 0$.

2. The Divisor of an Automorphic Function

From now on we confine ourselves to groups Γ that have normal polygons with a finite number of sides. Since Γ has been assumed to be horocyclic from 1F on, the normal polygon has no free sides. We shall see later that such groups correspond to Riemann surfaces that are compact or can be compactified by the adjunction of a finite number of points.

If f is a nonconstant automorphic function on Γ, it will have a certain number of zeros and poles in a normal polygon. The *divisor* of f, denoted by $D(f)$, is a formal symbol that states the position of zeros and poles of f together with their multiplicities. We shall have to make careful rules for counting the multiplicity of a zero or pole, especially at boundary points of the normal polygon. The *degree* of $D(f)$, written deg $D(f)$, is the sum of the multiplicities; that is, it is the number of zeros minus the number of poles of f in a normal polygon, counted in correct multiplicity. The fundamental theorem of this section is that

$$\deg D(f) = 0.$$

A consequence of this theorem is that f has a *valence*; that is, f assumes

† This implies that $F(x + iy) \to a$, say, as $y \to +\infty$, *uniformly in x*.

each complex value the same number of times in a normal polygon. The valence must therefore be the number of poles of f in the polygon, and we shall prove that this number is finite as well as positive.

2A. For the time being we confine ourselves to points in the interior of the normal polygons N_0, where a pole may be recognized from the ordinary Laurent expansion of f. If f has a pole at τ_0, $|f(\tau)| \to +\infty$ as $\tau \to \tau_0$. Hence $|f(\tau)| \to \infty$ as $\tau \to V\tau_0$ for each $V \in \Gamma$; that is, f has a pole at $V\tau_0$. Thus f has the same number of poles in the interior of every normal polygon.

THEOREM. A function $f \in \{\Gamma, 0\}$ has a finite number of zeros and poles in N_0.

As yet we do not assert that this number is positive.

Since $f = 0$ at a point if and only if $1/f$ has a pole there, and $1/f \in \{\Gamma, 0\}$, we need prove the theorem only for poles. If f has infinitely many poles in the normal polygon N_0, there will be a point τ^* where the poles accumulate. Clearly τ^* is a limit point of Γ—at an ordinary point f is meromorphic—but also $\tau^* \in \bar{N}_0$. Hence τ^* must be a point of E. Either τ^* is on a side of N_0 (but not on a free side†) or τ^* is on no side but its every neighborhood meets sides. Since N_0 has a finite number of sides, the second possibility is ruled out. Therefore τ^* is not a parabolic vertex.

We suppose as a first case that $\tau^* = i\infty$. Then N_0 lies in a strip

$$S : \xi < x < \xi + \lambda, \qquad y > y_0 > 0.$$

Let

$$t = e\left(\frac{\tau}{\lambda}\right), \qquad \tau = x + iy \in S. \tag{9}$$

Since $|t| = \exp(-2\pi y/\lambda)$, t maps S into a circle D with center $t = 0$, the point $\tau = i\infty$ going into the center. The reader is referred to the figure on page 79.

The sides s_1, s_2 go into coincident radii of D described in opposite directions and σ goes into D. If we let S^+ be S (open) together with s_1 and s_2, then t maps S^+ *onto* the interior of D in one-to-one fashion except that the points $z_1 \in s_1$, $z_2 \in s_2$, with $z_2 = z_1 + \lambda$, go into the same point of the radius OA.

If we now define

$$f(\tau) = \mathfrak{f}(t),$$

$\mathfrak{f}$ is meromorphic in D deleted by the radius OA. Indeed, $\mathfrak{f}$ is single-valued in this region, and from

$$f'(\tau) = \mathfrak{f}'(t) \cdot \frac{dt}{d\tau} = \frac{2\pi i}{\lambda}\, t\mathfrak{f}'(t)$$

† A free side consists of ordinary points. Besides, we are considering only horocyclic groups.

we deduce that $\hat{f}$ is regular at $t = e(\tau/\lambda)$ if f is regular at τ. Likewise $\hat{f}$ has a pole at t_0 if f has a pole at τ_0, since $|f(\tau)| \to \infty$ as $\tau \to \tau_0$ and so $|\hat{f}(t)| \to \infty$ as $t \to t_0$. Moreover, $\hat{f}(t)$ tends to the same value as t approaches a point of OA from opposite sides. For this corresponds to $\tau \to iy$, $\tau \to iy + \lambda$, and $f(\tau) = f(\tau + \lambda)$. *Thus $\hat{f}$ is meromorphic in* U *except possibly at* $t = 0$.

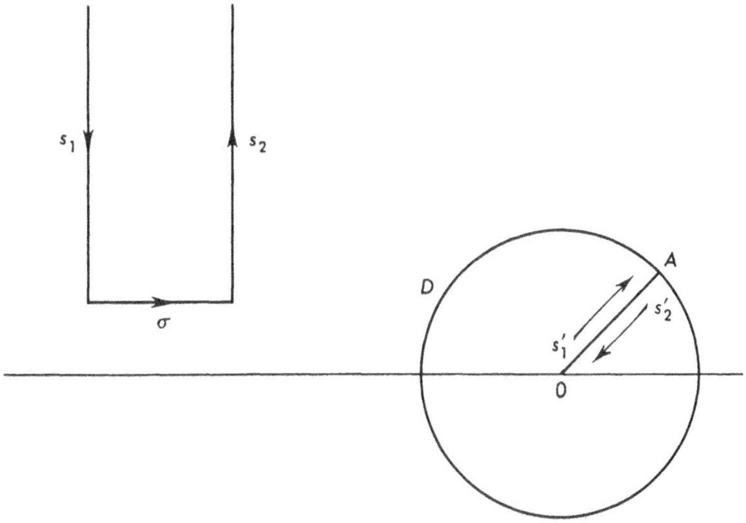

FIG. 21.

LEMMA. Let $g(t)$ be regular in $0 < |t| < \rho, \rho > 0$, except for poles at $t = t_n$, $n = 1, 2, \cdots$, where $t_n \to 0$. Then $g(t)$ approaches every complex number arbitrarily closely in every neighborhood of $t = 0$.

Let c be the complex number. If $c = \infty$, the result is trivial. Otherwise, define

$$h(t) = \frac{1}{g(t) - c}.$$

Then h is regular in $0 < |t| < \rho$, for we may assume $g(t) \neq c$, and $h(t_n) = 0$. Moreover, we may assume $|g(t) - c|$ bounded away from 0; that is, $h(t)$ is bounded. By Riemann's theorem, $h(t)$ may be defined so as to be regular at $t = 0$, and $h(t_n) = 0$ implies $h(t) \equiv 0$. This is a contradiction, since $g(t) = \infty$ only for $t = t_n$.

We return to the proof of the theorem. Supposing now that f has poles accumulating at $\tau = i\infty$, we select a subsequence of them: $\tau_n \to i\infty$. Hence $\hat{f}(t)$ has poles at the distinct points $t_n = e(\tau_n/\lambda)$ in U and $t_n \to 0$. It follows by the lemma that $\hat{f}$ does not tend to a definite limit as $t \to 0$, which means $f(\tau)$ tends to no definite limit for $\tau \to i\infty$. This contradicts the definition of $\{\Gamma, 0\}$.

The second and final case is that in which the poles accumulate at a finite parabolic vertex p. The idea of the proof is the same but we must use a different mapping function.

Let T be a parabolic sector constructed at p (see I, 4I); it consists of s images of a small curvilinear triangle in N_0 with apex at p, where s is the number of points in the cycle determined by p. If P generates the stability subgroup Γ_p, the sides s_1, s_2 of T are connected by P, $Ps_1 = s_2$. The remaining boundary of T,

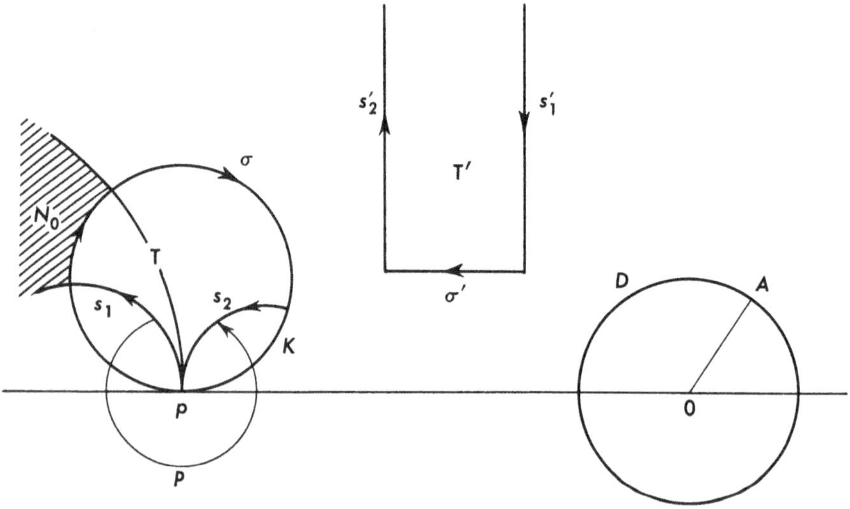

FIG. 22.

namely, σ, is an arc of K, a fixed circle of P. The sides s_1, s_2 are orthogonal to both E and σ. T is considered to include s_1, s_2 but not σ. We write P explicitly:

$$\tau' = P\tau, \qquad \frac{1}{\tau' - p} = \frac{1}{\tau - p} + c$$

where we may clearly assume $c > 0$.

Now consider the mapping

$$t = e\left(\frac{-1}{c(\tau - p)}\right). \tag{10}$$

We decompose this into two mappings:

$$\xi = \frac{-1}{c(\tau - p)}, \qquad t = e(\xi).$$

The first mapping is a linear transformation of determinant $c > 0$, which maps T onto a half-strip T′. By conformality s_1', s_2' are vertical lines and σ' is horizontal. T′ contains s_1', s_2' but not σ'. The point p is sent to $i\infty$. The function ξ is one-to-one from T onto T′. It is readily calculated that σ' has length 1. The second function $t = e(\xi)$ is similar to the one discussed previously. It maps T′ onto an open disk D centered at the origin.

Putting the two mappings together, we conclude that t maps T into D in a one-to-one manner except that points z_1, $z_2 = Pz_1$ in T are sent to the same t_1 of a certain radius OA of D.

If we now set

$$f(\tau) = \mathring{f}(t), \qquad t = e\left(\frac{-1}{c(\tau - p)}\right)$$

then by the reasoning used in the previous case $\mathring{f}$ is meromorphic in D. We obtain, exactly as before, a contradiction by assuming $f(\tau)$ has a sequence of poles converging to p.

This concludes the proof of the theorem. We have shown incidentally that $\mathring{f}(t)$ is meromorphic at $t = 0$, a fact we may express by saying that $f(\tau)$ is meromorphic at $\tau = p$ (which may be $i\infty$). This amounts, of course, to a definition of meromorphicity of f at a parabolic vertex: *we say $f(\tau)$ is meromorphic at a parabolic vertex $\tau = p$ if and only if $\mathring{f}(t) = f(\tau)$ is meromorphic at $t = 0$, where t is given by* (9) *if $p = i\infty$ and by* (10) *otherwise.* Defining H^+ to be the union of H and the set of parabolic vertices of Γ, we may restate the definition of an automorphic function as a Γ-invariant function that is meromorphic on H^+.

COROLLARY. A function $f \in \{\Gamma, 0\}$ has a finite number of zeros and poles in $\bar{N}_0$.

This follows from the fact that N_0 has a finite number of sides by assumption and so a finite number of parabolic vertices.

2B. We shall now prove the fundamental result that a nonconstant automorphic function must have at least one pole. We give two proofs.

FIRST PROOF. Let $f \in \{\Gamma, 0\}$ and let N be a fixed normal polygon of Γ. In contradiction to the theorem we suppose the least upper bound B of $|f(\tau)|$ in H is finite. There is then a sequence $\tau_n \in$ H such that $|f(\tau_n)| \to B$. We may assume that $\tau_n \in \bar{N}$ since f is invariant under Γ. The set $\{\tau_n\}$ has an accumulation point τ_0 lying in $\bar{N}$ and on a subsequence $\tau_p \to \tau_0$. If τ_0 is in H, then $|f(\tau_0)| = B$ by continuity, and $f =$ constant by the Maximum Modulus Principle applied to H.

Hence assume τ_0 belongs to E and so to $\bar{N} \cap$ E. Then τ_0 is a parabolic vertex p. Let K be the interior of a fixed circle through p. (If $p = i\infty$, K will be an upper half-plane.) As we have seen in 2A, the function t, given by (9) when $p = i\infty$ and by (10) otherwise, maps K into a disk D in U and in the mapping p goes to 0. From $f(\tau) = \mathring{f}(t)$ we conclude that $\mathring{f}$ assumes its maximum in D at the interior point 0. Since $\mathring{f}$ is regular, it follows that it is constant; the same, therefore, is true of f.

SECOND PROOF. If N is relatively compact in H, the result follows immediately from the Maximum Modulus Principle. Otherwise denote the parabolic vertices in N by $p_1, \cdots, p_s$. Since f approaches a limit on vertical approach to p_i, we may write

$$f_i = \lim_{\tau \to p_i} f(\tau).$$

Consider

$$F(\tau) = \prod_{i=1}^{s} (f(\tau) - f_i).$$

Clearly $F \in \{\Gamma, 0\}$ and F is bounded in H; moreover,

$$\lim_{\tau \to p_i} F(\tau) = 0, \qquad i = 1, \cdots, s. \tag{11}$$

Let C be the least upper bound of $|F(\tau)|$ in H. Because of (11) there is a region $N^* \subset N$ (obtained, for example, by truncating N slightly) such that $\bar{N}^* \subset$ H and $|F(\tau)| < C/2$ in $N - N^*$. Hence the maximum modulus of F in H cannot be assumed in $N - N^*$ and so must be assumed at a point of H. This shows $F =$ constant and then (11) shows $F \equiv 0$. Hence $f(\tau) \equiv f_i$ for some i.

We can state our result as follows.

THEOREM. An automorphic function that is bounded in a fundamental region is constant.

For the function is bounded in the closure of the fundamental region and therefore in H.

Exercise 1. Let Γ be a group of the second kind; that is, Γ is not horocyclic. We say $f \in \{\Gamma, 0\}$ if f is Γ-invariant and meromorphic on $\mathscr{O}^+ = \mathscr{O} \cup$ P. (At a parabolic vertex we require f to approach limits as $\tau \to p$ from within a fundamental region $N_0 \subset$ H and also as $\tau \to p$ with τ confined to the mirror image of N_0 in the lower half-plane; these limits need not be the same. Similarly for $p = \infty$.) Prove the above theorem for f automorphic on a group of the second kind.

2C. Before we can discuss divisors we must define what we mean by the order of a function $f \in \{\Gamma, 0\}$ at a point $\tau \in$ H$^+$.

If $\tau_0 \in$ H and is not a fixed point, we set $t = \tau - \tau_0$ and

$$f(\tau) = \hat{f}(t) = \sum_{h=\mu}^{\infty} a_h t^h, \qquad t = \tau - \tau_0, \qquad a_\mu \neq 0, \qquad \mu < \infty \tag{12}$$

This is simply the usual Laurent expansion of f. We call μ the order of f at τ_0 and write

$$n(\tau_0, f) = \mu.$$

As usual we say τ_0 is a zero of order μ if $\mu > 0$, τ_0 is a pole of order $-\mu$ if $\mu < 0$.

Next suppose τ_0 to be an elliptic fixed point of order $l \geq 2$. Then $\tau_0 \in \mathrm{H}$. The expansion (12) is still valid but we shall show it possesses a special property.

Let E generate the subgroup Γ_{τ_0}:

$$\tau' = E\tau : \frac{\tau' - \tau_0}{\tau' - \bar{\tau}_0} = \epsilon \frac{\tau - \tau_0}{\tau - \bar{\tau}_0}, \qquad \epsilon = e\left(\frac{k}{l}\right)$$

where $(k, l) = 1$. Define

$$t_1 = \frac{\tau - \tau_0}{\tau - \bar{\tau}_0}, \qquad t_1' = \frac{\tau' - \tau_0}{\tau' - \bar{\tau}_0}$$

so that E appears as

$$t_1' = \epsilon t_1 .$$

If we set

$$g(t_1) = f(\tau), \qquad g(t_1') = f(\tau'),$$

the invariance of f under $\tau \to \tau'$ gives $g(t_1') = g(t_1)$, or

$$g(t_1') = g(\epsilon t_1) = g(t_1).$$

Clearly g is meromorphic in a neighborhood of $t_1 = 0$, the point corresponding to $\tau = \tau_0$. Introduce the Laurent series

$$g(t_1) = \sum_{h=\mu_1}^{\infty} b_h t_1^h, \qquad b_{\mu_1} \neq \infty, \qquad \mu_1 < \infty$$

into the above equation and get

$$b_h(\epsilon^h - 1) = 0.$$

It follows that only those coefficients survive for which h is a multiple of l. With a change of notation this gives

$$f(\tau) = \mathring{f}(t) = \sum_{n=\mu}^{\infty} a_n t^n, \qquad a_\mu \neq 0, \qquad \mu < \infty \tag{13}$$

where we have set

$$t = t_1^l = \left(\frac{\tau - \tau_0}{\tau - \bar{\tau}_0}\right)^l.$$

We define

$$n(\tau_0, f) = \mu.$$

At a parabolic vertex p the function $f(\tau) = \mathring{f}(t)$ is meromorphic at $t = 0$

(see end of 2A), where t is given by (9) when $p = i\infty$ and by (10) when p is finite. Hence we have the expansions:

$$\tau_0 = i\infty : f(\tau) = \mathring{f}(t) = \sum_{h=\mu}^{\infty} a_h t^h, \qquad a_\mu \neq 0, \qquad \mu < \infty, \qquad t = e\left(\frac{\tau}{\lambda}\right) \tag{14}$$

$$\tau_0 = p : f(\tau) = \mathring{f}(t) = \sum_{h=\mu}^{\infty} a_h t^h, \qquad a_\mu \neq 0, \qquad \mu < \infty, \qquad t = e\left(\frac{-1}{c(\tau - p)}\right) \tag{15}$$

In each case

$$n(\tau_0, f) = \mu.$$

We call (14) the *Fourier series* of f.

We have now defined the order of f at all points of H^+. The variable t occurring in (12)—(15) is called the *local variable* or *local uniformizer*. The order function $n(\tau_0, f)$ has been defined for all $f \not\equiv 0$, and for $f \equiv 0$ we set

$$n(\tau_0, 0) = +\infty.$$

We observe that

$$\begin{aligned} n(\tau, fg) &= n(\tau, f) + n(\tau, g), \\ n(\tau, f + g) &\geq \min \{n(\tau, f), n(\tau, g)\}. \end{aligned} \tag{16}$$

We now define the *valence* $N(f)$ as

$$N(f) = \sum_{\substack{n(\tau,f)<0 \\ \tau \in N^*}} n(\tau, f),$$

where N^* is a fundamental set of Γ (see I, 4A). (A fundamental set may be obtained from a normal polygon by adjoining one point from each cycle and one open side from each pair of conjugate sides.) $N(f)$ is the number of poles of f in N^*, each counted according to its multiplicity. We shall see immediately that $N(f)$ is independent of the fundamental set chosen.

Exercise 1. $f \in \{\Gamma, 0\}$ is never a trigonometric polynomial (that is, its Fourier series is never right-finite).
[Use 1A, Exercise 1.]

2D. An automorphic function f has the same order at equivalent points:

$$n(\tau, f) = n(V\tau, f), \qquad V \in \Gamma, \qquad \tau \in H^+.$$

In order to prove it let us remember that $f \not\equiv 0$ has an expansion in a local variable t at every point of H^+,

$$f(\tau) = \mathring{f}(t) = \sum_{h=\mu}^{\infty} a_h t^h, \qquad a_\mu \neq 0.$$

(If $f \equiv 0$, the result we are after is obvious.) Hence

$$\mu = \frac{1}{2\pi i}\int_C d\log \hat{f}(t),$$

where C is a euclidean circle centered at the origin and containing no zero or pole of $\hat{f}$ other than $t = 0$.

Suppose $\tau_2 = V\tau_1$, $\tau' = V\tau$ with $V \in \Gamma$. Let t, t' be the local variables corresponding to τ_1, τ_2, respectively. Then

$$\mu' = \frac{1}{2\pi i}\int d\log \hat{f}(t'),$$

the integration being over a circle defined similarly to C. We now write $t = t(\tau)$ and select a single-valued branch of the inverse function $\tau = \tau(t)$; set $\sigma = \tau(C)$, $\sigma' = V\sigma$, $C' = t'(\sigma')$. Using the invariance of $d \log f(\tau)$ under Γ we get

$$\mu = \frac{1}{2\pi i}\int_C d\log \hat{f}(t) = \frac{1}{2\pi i}\int_\sigma d\log f(\tau) = \frac{1}{2\pi i}\int_{\sigma'} d\log f(V^{-1}\tau)$$

$$= \frac{1}{2\pi i}\int_{\sigma'} d\log f(\tau) = \frac{1}{2\pi i}\int_{C'} d\log \hat{f}(t') \tag{17}$$

To show $\mu = \mu'$, then, it is sufficient to show C' is a euclidean circle about $t' = 0$; for by taking C small enough we can be sure that there are no zeros or poles of $\hat{f}$ other than $t' = 0$.

We now get down to cases.

(1) $\tau_1 \in \mathrm{H}$ and is not a fixed point. Here $t = \tau - \tau_1$, $t' = V\tau - V\tau_1$, and the result is immediate.

(2) $\tau_1 \in \mathrm{H}$ and is an elliptic vertex of order l. Here $t = (\tau - \tau_1)^l(\tau - \bar{\tau}_1)^{-l}$. Consider $t_1 = (\tau - \tau_1)(\tau - \bar{\tau}_1)^{-1}$; this is a nonsingular linear transformation whose inverse $t_1 \to \tau$ maps C on a circle K enclosing τ_1. Since C is orthogonal to every ray from the origin—that is, every line connecting 0 and ∞—K will be orthogonal to every circle connecting τ_1 and $\bar{\tau}_1$ and so will be an H-circle about τ_1 (I, 4B, Exercise 2). The figure consisting of C and one of its radii described twice in opposite directions is mapped by $\tau(t)$ onto a curvilinear triangle S consisting of an arc σ of K plus two circular arcs meeting at τ_1 (see Figure 23). (Actually S is an elliptic sector.) By conformality the angle at τ_1 will be $2\pi/l$. Clearly σ is mapped by V onto an arc σ' having the same relation to $V\tau_1$ as σ has to τ_1. Finally $t' = (V\tau - V\tau_1)^l(V\tau - \overline{V\tau_1})^{-l}$ maps σ' on a full circle C'.

(3) τ_1, τ_2 are finite parabolic vertices. Referring to Figure 22 and relevant discussion, we find that σ is an arc subtending a parabolic sector at τ_1. Hence σ' is a similar arc subtending a parabolic sector at τ_2, and so, reversing the mapping, we deduce that C' is a euclidean circle about $t' = 0$.

(4) $\tau_1 = i\infty$, $\tau_2 =$ finite parabolic vertex. From C to σ we use the mapping inverse to $t = e(\tau/\lambda)$; see Figure 21 and the relevant discussion. Then σ is a horizontal line segment of length λ. But σ' is a circle tangent to E at τ_2, and the previous discussion holds: C' is a euclidean circle.

This completes the proof of the

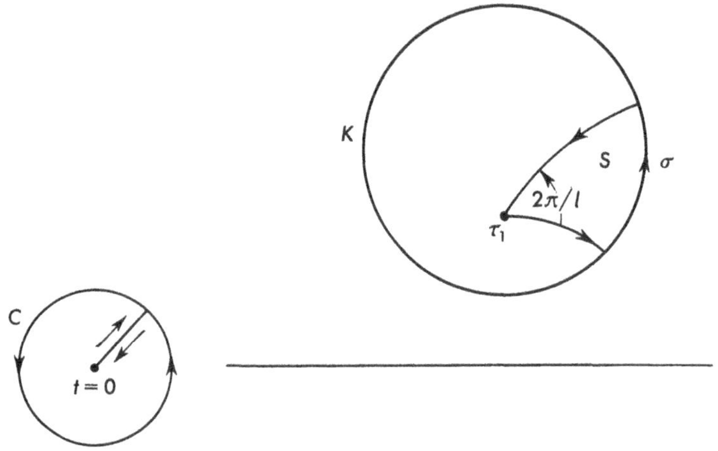

FIG. 23.

THEOREM. For $f \in \{\Gamma, 0\}$, $V \in \Gamma$, and $\tau \in H^+$, we have

$$n(V\tau, f) = n(\tau, f).$$

2E. We recall that N^* is a fundamental set for Γ (relative to H^+) if it contains exactly one point from each orbit

$$\Gamma x, \qquad x \in H^+.$$

By the theorem just proved, VN^* is a fundamental set for each $V \in \Gamma$. Let f be an automorphic function on Γ and let $f \not\equiv 0$. Since the zeros and poles of f are the same in each fundamental set, we need to specify them only in one N^*. A divisor is a formal symbol that does just that.

We define a *divisor* as a formal (finite) sum

$$D = \sum_{\tau \in N^*} u(\tau) \cdot \tau,$$

where $u(\tau)$ is a rational number, $u(\tau) = 0$ except for a finite number of τ, and

$$u(V\tau) = u(\tau), \qquad V \in \Gamma. \tag{18}$$

Because of the last condition D is independent of the particular fundamental set N^* used in its definition. We say D has order $u(\tau)$ at τ. The divisor 0 is obtained by setting $u(\tau) \equiv 0$.

When we choose $u(\tau)$ to be the order of f at τ, D becomes $D(f)$, *the divisor of the function* f:

$$D(f) = \sum_{\tau \in N^*} n(\tau, f) \cdot \tau.$$

Since $n(\tau, f) = 0$ except where f has a zero or pole and (by Theorem 2A) f has only a finite number of zeros and poles in N^*, we see that D is a finite sum. Besides this, $n(\tau, f)$ satisfies (18) by Theorem 2D and it is rational, as we saw in 2C. Hence $D(f)$ is a divisor.

A divisor that is the divisor of a function is called a *principal divisor.* The divisor 0 is $D(1)$ and so is a principal divisor.

By defining

$$D_1 - D_2 = \sum u^{(1)}(\tau) \cdot \tau - \sum u^{(2)}(\tau) \cdot \tau = \sum (u^{(1)}(\tau) - u^{(2)}(\tau)) \cdot \tau,$$

we can make an additive abelian group from the set of all divisors; the zero element is the divisor 0. The set of all principal divisors forms a subgroup, for $D(f_1) - D(f_2) = D(f_1/f_2)$. The elements of the factor group are called *divisor classes.* Thus a divisor class consists of all divisors differing from a given divisor by a principal divisor.

The *degree of a divisor* is

$$\deg D = \sum_{\tau \in N^*} u(\tau),$$

so that

$$\deg D(f) = \sum_{\tau \in N^*} n(\tau, f).$$

The object of this section is to prove that *the degree of a principal divisor is zero*[†]:

$$\deg D(f) = 0.$$

This will show, in particular, that all divisors in the same divisor class have the same degree, called the degree of the divisor class, since any two of them differ by a principal divisor.

Write

$$J = \deg D(f).$$

We shall treat J by contour integration. Let S be the boundary of N_0. We make a preliminary modification of S in case there is an interior point τ_1 of a side s where the function f has a zero or pole. We deform S by a circular arc around τ_1 and by an equivalent arc around the point τ_2 equivalent to τ_1 on the conjugate side s'. Next, if $\{\tau_1, \cdots, \tau_h\}$ is an accidental cycle on S, a small circle K around τ_1 can be divided into h arcs; the arcs outside N_0 are equivalent under Γ to

[†] The converse, however, is false; that is, not every divisor whose degree is 0 is the divisor of an automorphic function.

arcs C_j at τ_j, $j = 2, \cdots, h$. We deform S around the C_j, $j > 1$, but at τ_1 we make S describe K so as to include τ_1. We still call the deformed contour S and its interior N_0. Then N_0 is a fundamental region, its boundary S consists of pairs of conjugate sides, and there are no zeros or poles of f on S except the ones, if any, at elliptic or parabolic vertices.

At an elliptic or parabolic cycle $\{\tau_1, \cdots, \tau_h\}$ we make the following detours, obtaining a new contour S' bounding a region N_0'. A fixed circle of an elliptic (parabolic) generator of the subgroup Γ_{τ_1} contains an arc σ of an elliptic (parabolic) sector and we can choose σ so that it crosses a block of h consecutive normal polygons with vertex τ_1, one of which is N_0. Let C_1 be the part of σ lying in N_0; the remaining pieces of σ, say C_j', $j = 2, \cdots, h$, are equivalent by $T_j \in \Gamma$ to arcs $C_j = T_j C_j'$ around τ_j. The detours are $\{C_j, j = 1, \cdots, h\}$.

In each case we make the detours small enough so that they contain no zero or pole except the possible one at the point in question. Only a finite number of detours are required. The region N_0' includes all points τ of a fundamental set at which $n(\tau, f) \neq 0$ except possibly for elliptic and parabolic vertices.

If we write Δ for the union of the detours (around elliptic and parabolic vertices) and set $S^* = S' - \Delta$, then by a classical theorem of function theory we have

$$J - \sum_{v} n(\tau, f) = \frac{1}{2\pi i} \int_{S^*} d \log f(\tau) + \frac{1}{2\pi i} \int_{\Delta} d \log f(\tau), \tag{19}$$

the sum being extended over a system of elliptic and parabolic vertices of N_0, one from each cycle.

Suppose $\{\tau_1, \cdots, \tau_h\}$ is an elliptic cycle. From the discussion in 2D we have

$$n(\tau_1, f) = \frac{1}{2\pi i} \int_{\sigma} d \log f(\tau);$$

where σ is the arc of an elliptic sector at τ_1. On the other hand $T_j^{-1} C_j = C_j'$, $j = 2, \cdots, h$, and C_1 and the arcs $\{C_j'\}$ fit together to make up an arc σ. Because of the invariance of $d \log f$ under $\tau \to T_j \tau$, we get

$$n(\tau_1, f) = -\sum_{j=1}^{h} \frac{1}{2\pi i} \int_{C_j} d \log f(\tau).$$

The minus sign arises because σ is described in the negative sense (region to the right), whereas C_j is described positively.

Exactly the same argument applies to a parabolic cycle, and we conclude that

$$\frac{1}{2\pi i} \int_{\Delta} d \log f(\tau) = -\sum_{v} n(\tau, f);$$

then (19) becomes

$$J = \frac{1}{2\pi i} \int_{S^*} d \log f(\tau).$$

As the detours shrink to their points, J does not change, because of the requirement that the detours enclose only the zero or pole at the vertex. Hence

$$J = \frac{1}{2\pi i}\int_S d\log f(\tau).$$

The boundary of S consists of an even number of sides in conjugate pairs. However, the contributions from the sides in a conjugate pair cancel. Indeed, if s and $s' = Vs$ form a pair, then $d\log f(\tau)$ is unchanged under $\tau \to V\tau$; moreover, s' is described negatively if s is described positively. This concludes the proof of the following

THEOREM. Let Γ be a horocyclic group having a normal polygon with a finite number of sides. Let $f \in \{\Gamma, 0\}$ and suppose $f \not\equiv 0$. Then

$$\deg D(f) = 0.$$

Exercise 1. Prove Theorem 2B by means of the above theorem.

Exercise 2. Carry out the above proof for a group that is not horocyclic.

2F. The theorem announced at the beginning of this section now follows easily. We recall that $N(f)$ is the number of poles of f in N^*, counted in correct multiplicity.

THEOREM. A nonconstant automorphic function assumes each complex value $N(f)$ times in any fundamental set N^*.

Since f is not constant, $N(f)$ is positive. Now $f - c$ is automorphic and has $N(f)$ poles in N^*; it therefore has $N(f)$ zeros. But $f - c$ is zero at exactly those points where f assumes the value c.

This theorem shows that an automorphic function is completely determined by the principal parts of its expansions at the poles in N^* and its value at one other point. For the difference of two functions with the same principal parts is regular in N^* and so is a constant.

2G. We have seen that a nonconstant automorphic function has a positive valence. The lowest possible valence is 1. But are there actually univalent functions (functions of valence 1)—that is, functions that assume each value exactly once in the fundamental region? This depends entirely on the group Γ and, in fact, on a topological characteristic of Γ called the genus. We shall see in the next chapter that groups of genus 0, and only those groups, support univalent functions. It is noteworthy that this question, which can be stated entirely in the context of analytic functions, is settled completely by reference to a purely topological quantity.

2H. THEOREM. If $f \in \{\Gamma, 0\}$, $g \in \{\Gamma, 0\}$, then f, g satisfy an algebraic equation

$$P(f, g) = 0$$

with complex coefficients.

Let

$$\Phi(\tau) = \sum_{i=0}^{s} \sum_{j=0}^{t} a_{ij} f^i(\tau) g^j(\tau),$$

with arbitrary complex a_{ij} and arbitrary integers $s, t \geq 0$. Φ is automorphic on Γ. We shall show that constants a_{ij} can be chosen so that Φ has more zeros than poles in a normal polygon. Then $\Phi \equiv 0$ by Theorem 2E and this establishes the result.

Suppose $N(f) = n$, $N(g) = m$. Then we have the inequality

$$N(\Phi) \leq ns + mt. \tag{20}$$

Let

$$\nu = (s + 1)(t + 1) - 1;$$

$\nu + 1$ is the number of constants in Φ. Select ν points $\tau_1, \cdots, \tau_\nu$ in a fixed normal polygon N_0 that are distinct from each other and from the poles of f and g. The conditions

$$\Phi(\tau_1) = 0, \cdots, \qquad \Phi(\tau_\nu) = 0$$

are equivalent to ν linear equations in $\nu + 1$ unknowns $\{a_{ij}\}$; this system, therefore, always has a nontrivial solution. The polynomial Φ formed with such constants has at least ν zeros in N_0. But for s and t large enough

$$\nu = (s + 1)(t + 1) - 1 > ns + mt. \tag{21}$$

With (20) this shows that Φ has more zeros than poles and so $\Phi \equiv 0$.

The theorem is not necessarily true if we relax the condition that N_0 has a finite number of sides or the condition that f tends to a limit at a parabolic vertex. Thus e^f, e^g are not necessarily connected by an algebraic relation.

2I. Suppose f is univalent in N_0; that is, $n = 1$. Then (21) is satisfied for $s = m$, $t = 1$. Hence

$$g\Omega_1(f) + Q_2(f) = 0,$$

where Q_1, Q_2 are polynomials. If $Q_1 \equiv 0$, we would have $Q_2(f) \equiv 0$, which implies that f assumes only a finite number of values—namely, the zeros of Q_2.

Then by continuity f is a constant, contrary to the hypothesis that f is univalent. Hence $Q_1 \not\equiv 0$ and we get

$$g(\tau) = \frac{-Q_2(f)}{Q_1(f)}.$$

That is, every automorphic function on Γ is a rational function of the univalent function f. Recalling the notation $\mathbf{K}(\Gamma)$ for the field of automorphic functions on Γ, we now have:

THEOREM. $\mathbf{K}(\Gamma)$ is the field of rational functions of f if f is a univalent function on Γ.

In the next chapter we shall see that whether Γ supports a univalent function or not, $\mathbf{K}(\Gamma)$ is isomorphic to a field of *algebraic* functions of one variable.

3. The Divisor of an Automorphic Form

Divisors can be introduced for automorphic forms as well as functions. Let $F \in \{\Gamma, -r\}$, $r =$ even integer. The first step is to find expansions for F at the points of H^+.

3A. Let us consider a special case: $F = \varphi'$, where $\varphi \in \{\Gamma, 0\}$. Since

$$\varphi'(V\tau)\, dV\tau = \varphi'(\tau)\, d\tau,$$

it is clear that $\varphi' \in \{\Gamma, -2\}$. The expansions for φ' are, of course, obtained by simply differentiating those for φ. We have

$$\varphi(\tau) = \sum_{h=\mu}^{\infty} a_h t^h = \hat{\varphi}(t),$$

where t is the appropriate local variable (2C), and so

$$\varphi'(\tau) = \frac{dt}{d\tau} \cdot \sum_{h=\mu}^{\infty} h a_h t^{h-1}.$$

There are, as usual, four cases.

(1) $\tau_0 \in H$, nonfixed: $dt/d\tau = 1$

(2) τ_0 elliptic fixed point of order l:

$$\frac{dt}{d\tau} = l \left(\frac{\tau - \tau_0}{\tau - \bar{\tau}_0}\right)^{l-1} \frac{\tau_0 - \bar{\tau}_0}{(\tau - \bar{\tau}_0)^2},$$

$$(\tau - \bar{\tau}_0)^2\, dt/d\tau = C t^{1-1/l}, \qquad C \neq 0$$

(3) $\tau_0 = i\infty$ (parabolic vertex):

$$\frac{dt}{d\tau} = \left(\frac{2\pi i}{\lambda}\right) e\left(\frac{\tau}{\lambda}\right) = Ct, \qquad C \neq 0$$

(4) $\tau_0 =$ finite parabolic vertex:

$$(\tau - \tau_0)^2 \left(\frac{dt}{d\tau}\right) = \frac{2\pi i t}{c} = Ct, \qquad C \neq 0.$$

In all cases, therefore, the local variable t is the same as before, but in (2) and (4) *the function that is expanded changes.*

We can expand $F = (\varphi')^m$ by merely taking the mth power. Now $(\varphi')^m \in \{\Gamma, -2m\}$. Also if $H \in \{\Gamma, -2m\}$, then

$$\frac{H}{(\varphi')^m} = f \in \{\Gamma, 0\};$$

hence $H = f(\varphi')^m$ and the expansion of H is seen to be of the same form as the expansion of $(\varphi')^m$. If we had used another function $(\psi')^m$ instead of $(\varphi')^m$, we should have

$$\frac{H}{(\psi')^m} = g \in \{\Gamma, 0\};$$

since the expansions of ψ' and g are of the same form as those of φ' and f, respectively, the expansion of H would not change. We have proved:

THEOREM. If $F \in \{\Gamma, -r\}$, $r =$ even integer, F has the following expansions:

(1) $$F(\tau) = \hat{F}(t) = \sum_{h=\mu}^{\infty} a_h t^h, \qquad t = \tau - \tau_0; \qquad \tau_0 \in \mathrm{H}, \text{ nonfixed}$$

(2) $$(\tau - \bar{\tau}_0)^r F(\tau) = \hat{F}(t) = \sum_{h=\mu}^{\infty} a_h t^{h-(r/2l)}, \qquad t = \left(\frac{\tau - \tau_0}{\tau - \bar{\tau}_0}\right);$$

τ_0 is elliptic, order l

(3) $$F(\tau) = \hat{F}(t) = \sum_{h=\mu}^{\infty} a_h t^h, \qquad t = e\left(\frac{\tau}{\lambda}\right); \qquad \tau_0 = i\infty, \text{ parabolic}$$

(4) $$(\tau - \tau_0)^r F(\tau) = \hat{F}(t) = \sum_{h=\mu}^{\infty} a_h t^h, \qquad t = e\left(\frac{-1}{c(\tau - \tau_0)}\right),$$

$\tau_0 =$ finite parabolic vertex.

We define $n(\tau, F) = \mu$ except in case (2) when $n(\tau, F) = \mu - r/2l$.

The rules (16) for $n(\tau, f)$ hold also when f, g are automorphic forms.

Exercise 1. For p a parabolic vertex (finite or infinite) we define $L(p)$ to be 0, a_0, or ∞ according as $\mu > 0$, $\mu = 0$, or $\mu < 0$. Show that for certain positive constants α, β,

$$F(\tau) - L(i\infty) = O(e^{-\alpha y}), \qquad p = i\infty$$

$$(\tau - p)^r F(\tau) - L(p) = O(e^{-\beta/y}), \qquad p \neq i\infty$$

as $\tau = x + iy \to p$ from within a normal polygon. In words, an automorphic form assumes its limiting value at a parabolic cusp *with exponential rapidity.* [A Stolz angle A is the interior of a V-shaped region with apex at p (the interior of a vertical strip if $p = \infty$). Since A includes the region formed by two sides of a normal polygon meeting at p, it is enough to consider approach to p within A.]

Exercise 2. $F \in \{\Gamma, -r\}$ has a zero or pole at the fixed point of an elliptic transformation of order l unless l divides $r/2$.

3B. THEOREM. An automorphic form has only a finite number of zeros and poles in any fundamental set N^*.

We can confine ourselves to the case of poles since $1/F$ belongs to $\{\Gamma, r\}$ if $F \in \{\Gamma, -r\}$. We follow the proof of Theorem 2A. An infinite set of poles of F has an accumulation point τ^* that is a parabolic vertex and lies on E. If $\tau^* = i\infty$, $F(\tau) = \hat{F}(t)$ with $t = e(\tau/\lambda)$ has infinitely many poles near $t = 0$ and is not meromorphic there. If τ^* is finite, we apply the same argument to $(\tau - \tau^*)^r F(\tau) = \hat{F}(t)$.

3C. The order function $n(\tau, F)$ is invariant under Γ. Suppose $F = \varphi' \in \{\Gamma, -2\}$, $\tau_1 \in \mathrm{H}^+$, $\tau_2 = V\tau_1$, $V \in \Gamma$. Let t, t' be local variables at τ_1, τ_2, respectively. From

$$\varphi(\tau) = \hat{\varphi}_1(t) = \hat{\varphi}_2(t'),$$

we get

$$\varphi'(\tau) = \hat{\varphi}_1'(t) \cdot \left(\frac{dt}{d\tau}\right) = \hat{\varphi}_2'(t') \cdot \left(\frac{dt'}{d\tau}\right),$$

and $\hat{\varphi}_1'(t)$ has the same order in t as $\hat{\varphi}_2'(t')$ has in t'. From the calculations of 3A we see that $dt/d\tau$ (Cases 1 and 3), $(\tau - \bar{\tau}_1)^2 dt/d\tau$ (Case 2), and $(\tau - \tau_1)^2 dt/d\tau$ (Case 4) have the same order in t as the corresponding expressions with τ_1 replaced by τ_2 have in t'. This proves that $n(\tau, \varphi') = n(V\tau, \varphi')$.

In general $F \in \{\Gamma, -r\}$ can be written in the form $F = (\varphi')^{r/2} g$, where $\varphi, g \in \{\Gamma, 0\}$. Then $n(\tau, F) = (r/2)\, n(\tau, \varphi') + n(\tau, g)$ and the result follows.

THEOREM. For $F \in \{\Gamma, -r\}$, $V \in \Gamma$, and $\tau \in H^+$, we have

$$n(V\tau, F) = n(\tau, F).$$

3D. We can now define the *divisor of an automorphic form* F:

$$D(F) = \sum_{\tau \in N^*} n(\tau, F) \cdot \tau.$$

By Theorem 3A, $n(\tau, F)$ is a rational number; by Theorem 3B, D is a finite sum; by Theorem 3C, $n(\tau, F)$ is Γ-invariant. Therefore D is a divisor.

Two forms of the same dimension lie in the same divisor class (2E), for their quotient is an automorphic function and its divisor is principal:

$$D\left(\frac{F_1}{F_2}\right) = D(F_1) - D(F_2) = D(f),$$

if $F_1, F_2 \in \{\Gamma, -r\}$ and therefore $f = F_1/F_2 \in \{\Gamma, 0\}$. Denote by $\mathbf{D}_r$ the set of divisors of all forms of dimension $-r$. $\mathbf{D}_r$ is a coset of $\mathbf{D}/\mathbf{D}_0$. $\mathbf{D}_2$ is called the *canonical class* and is denoted by $\mathbf{Z}$.

We define $\deg D(F)$ as before; it depends only on the dimension $-r$ of F. Now $\deg D(F^k) = k \deg D(F)$. And if $F \in \{\Gamma, -r\}$, then $F = G^{r/2}g$, where G is a fixed but arbitrary member of $\{\Gamma, -2\}$ and $g \in \{\Gamma, 0\}$. Hence

$$\deg D(F) = \frac{r}{2} \deg D(G) + \deg g = \frac{r}{2} \deg D(G) = \frac{r}{2} J,$$

where

$$J = \deg D(G)$$

depends at most on the group Γ. We shall now evaluate J.

To do so we shall integrate $d \log G$ around a normal polygon N_0, the particular polygon chosen being of no consequence. We merely repeat the earlier discussion with the necessary modifications (see end of 2E).

Suppose $\{\tau_1, ..., \tau_h\}$ is an elliptic or parabolic cycle. Since G is of dimension -2, we have

$$d \log G(T_j\tau) = \frac{2d\tau}{\tau + d_j/c_j} + d \log G(\tau), \qquad j = 2, 3, ..., h$$

where $C_j = T_jC_j'$. Now $(\tau + d_j/c_j)^{-1}$ is bounded in the neighborhood of τ_1. This is because $-d_j/c_j = T_j^{-1}(\infty) \neq \tau_1$, since T_j^{-1} maps a finite point (namely τ_j) on τ_1. Hence

$$\int_{C_j'} \frac{d\tau}{\tau + d_j/c_j} \to 0$$

with $C_j' \to 0$, $j = 2, ..., h$. It follows that, just as in 2E, the contribution from the detour around a point is the negative of the order of G at that point.

By the same reasoning as before we now have

$$J = \deg D(G) = \frac{1}{2\pi i} \sum_{i=-n}^{n}{}' \int_{s_i} d \log G(\zeta),$$

where the boundary of N_0 consists of the conjugate pairs of sides $\{s_i, s_{-i}, i = 1, \cdots, n\}$, and Σ' means $i \neq 0$. From $G \in \{\Gamma, -2\}$, that is, $G(V\zeta)\, dV\zeta = G(\zeta)\, d\zeta$, we deduce

$$d \log G(V\zeta) = d \log G(\zeta) - d \log \frac{dV\zeta}{d\zeta}, \qquad V \in \Gamma;$$

hence, setting $\zeta' = V_i\zeta$, where $V_i s_i = s_{-i}$, we get

$$\int_{s_{-i}} d \log G(\zeta') = -\int_{s_i} d \log G(V_i\zeta)$$

$$= -\int_{s_i} d \log G(\zeta) + \int_{s_i} d \log V_i'(\zeta),$$

or

$$J = \frac{1}{2\pi i} \sum_{i=-n}^{n}{}' \int_{s_i} d \log V_i'(\zeta).$$

Let the arc s_i be parameterized by u, $0 \leq u \leq 1$, so that

$$V_i'(\zeta) = \frac{V_i'(u)}{\zeta'(u)} = \frac{w'}{\zeta'},$$

where we have set $\omega = V(u)$. Note that $\omega' = V_i'(u) = (cu + d)^{-2}$ is positive since u is real. Hence

$$\int_{s_i} d \log V_i'(\zeta) = -\int_0^1 d \log \omega'(u) - \int_0^1 d \log \zeta'(u).$$

As $\zeta(u)$ traverses s_i in the positive sense, ω traverses s_{-i} in the negative sense. Moreover, J is real. Therefore

$$J = \operatorname{Re} J = \frac{-1}{2\pi} \sum_{i=1}^{n}{}' \left\{ \int_0^1 d \arg \omega'(u) + \int_0^1 d \arg \zeta'(u) \right\}$$

$$= -\frac{1}{2\pi} \sum_{i=1}^{n}{}' \left\{ \Delta_{s_{-i}} \arg + \Delta_{s_i} \arg \right\} = -\frac{1}{2\pi} \sum_{i=-n}^{n}{}' \Delta_{s_i} \arg,$$

where $\Delta_{s_i} \arg$ is the change in argument of the tangent as it traverses s_i in the positive sense.

We see that $-2\pi J$ is the sum of the changes in the argument of the tangent over the sides of N_0. Since N_0 is simply connected, the total change in the argument of the tangent over the boundary of N_0 is 2π. This total change is made up of the changes over the sides, that is, $-2\pi J$, plus the jumps in the argument of the tangent at the vertices. From the figure it is seen that at a vertex of angle α the jump is $\pi - \alpha$. Thus

$$2\pi = -2\pi J + \sum_{i=-n}^{n}{}' (\pi - \alpha_i).$$

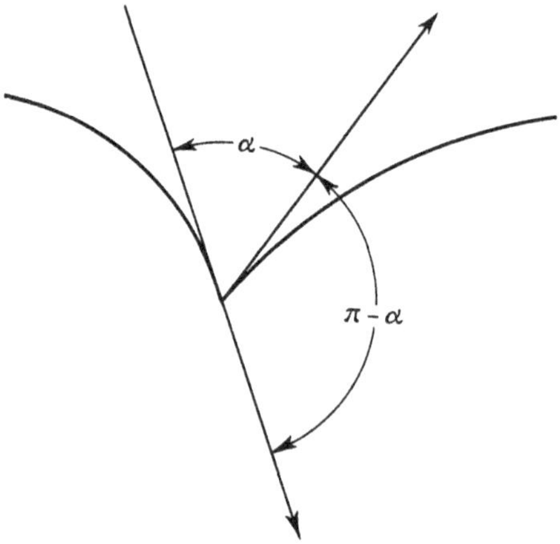

FIG. 24.

We shall calculate $\sum \alpha_i$ by considering the cycles of N_0. The sum of the angles at a cycle is $2\pi/l$, where l is defined on page 51. Hence

$$J = n - 1 - \sum \frac{1}{l}, \tag{22}$$

the sum in the right member being extended over all cycles of N_0. Combining this with equation (23) of I, 5C for the hyperbolic area $|N_0|$ of the normal polygon, we get

$$J = \deg D(G) = \frac{|N_0|}{2\pi}.$$

Hence

$$\deg D(F) = \frac{rJ}{2} = \frac{|N_0|}{4\pi}.$$

Using I, 5C, (26) to express $|N_0|$ in terms of the genus g, we finally get:

THEOREM. If $F \in \{\Gamma, -r\}$ we have

$$\deg D(F) = \frac{r}{2}\left\{2g - 2 + \sum_i (1 - l_i^{-1})\right\},$$

the sum being extended over the elliptic and parabolic cycles of Γ.

The theorem shows incidentally that the area of all normal polygons is the same, a fact we proved before (I, 5B).

3E. One consequence of the above is the

THEOREM. A form of positive dimension without poles is identically zero.

Positive dimension means $r < 0$, which implies $\deg D(F) < 0$. Thus there are more poles than zeros, so there must be at least one pole.

A form of nonnegative dimension cannot be everywhere regular without being a constant. This is not true of forms of negative dimension. Indeed the Poincaré series $G_{-r}(\tau, 0)$ provide a counterexample (see Theorem 1C). The series $G_{-r}(\tau, \nu)$ even vanish at the parabolic cusps for $\nu > 0$, yet they are not always constants. They are examples of *cusp forms* and we shall commence their study now. The principal problem is to determine when such series vanish identically.

Exercise 1. Let $C^+\{\Gamma, -r\}$ be the vector space of automorphic forms of dimension $-r$ that are regular at each point of H^+. Show from Theorem 3D that if $r > 0$, the dimension of $C^+\{\Gamma, -r\}$ is finite. (The dimension of a vector space $X \neq \{0\}$ is the lower bound of the number of elements in all subsets of X that span X; if $X = \{0\}$ we define its dimension to be zero.)

[Construct a linear combination of $\deg D(F) + 2$ linearly independent forms in C^+ that vanishes at $\deg D(F) + 1$ distinct points lying in the interior of one normal polygon.]

4. The Hilbert Space of Cusp Forms

The vector space $\{\Gamma, -r\}$ contains two important linear manifolds: $C^+\{\Gamma, -r\}$ consists of those $F \in \{\Gamma, -r\}$ that are regular at every point of H^+, while $C^0\{\Gamma, -r\}$ consists of those F in C^+ that vanish at each parabolic cusp of Γ. An element of C^0 is called a *cusp form*.

We have seen (Theorem 3E) that $C^+\{\Gamma, -r\}$ contains only 0 if $r < 0$ (that is, forms of *positive* dimension). Also $C^+\{\Gamma, 0\}$ contains only constants (Theorem 2B). *Hence we assume from now on that $r > 0$ and, as always, an even integer.*

By 3E, Exercise 1 the dimension of C^+, regarded as a vector space, is finite. Hence C^+ is a *subspace* of $\{\Gamma, -r\}$. It follows that C^0 is a subspace of C^+ and therefore of $\{\Gamma, -r\}$, and its dimension is finite.

The object of this section is to make C^0 into a *Hilbert* space and to discuss its properties. The Poincaré series which are cusp forms span C^0 and we shall consider the problem, not yet completely solved, of how to pick out a basis. The theory is due to H. Petersson.

4A. We recall that H is a Hilbert space if H is a linear space and if a scalar product (f, g) is defined for each pair $f, g \in H$. The scalar product is a complex-valued function that is assumed to be positive definite, hermitian symmetric, and bilinear. The norm of $f \in H$ is then taken to be $(f, f)^{1/2}$. Finally we require that H be complete in its norm (every Cauchy sequence converges to an element of H).

The last requirement is fulfilled automatically in spaces of finite dimension. Thus a vector space of finite dimension is a Hilbert space provided it has a scalar product.

Let N_0 be a normal polygon of Γ. For two complex-valued Lebesgue measurable functions f, g defined in H let

$$(f, g; N_0) = \iint_{N_0} f(\tau)\bar{g}(\tau)y^r \frac{dx\,dy}{y^2}, \qquad \tau = x + iy \tag{23}$$

where $\bar{g}(\tau)$ stands for the complex conjugate of $g(\tau)$. Assuming the integral converges, it is easily verified that $(f, g; N_0)$ has the properties of a scalar product. We shall now show that the integral does converge if f and g both belong to $C^+\{\Gamma, -r\}$ and at least one of them belongs to $C^0\{\Gamma, -r\}$, *provided* $r > 2$. (The condition $r > 2$ means, of course, $r \geqq 4$.)

4B. *From now on we assume*

$$r > 2.$$

We are going to prove the convergence of the integral (23). Let $f, g \in C^+\{\Gamma, -r\}$ while $f \in C^0\{\Gamma, -r\}$. We can partition N_0 into a compact subset S of H plus a finite number of triangular regions S_i, one at each finite parabolic cusp p_i, plus part of a vertical strip S_0 at $i\infty$ (see Figure 25). (In particular cases S_0 or the S_i may be absent.) The integral over S is clearly finite. We have from 3A, Exercise 1,

$$\left| \iint_{S_0} f(\tau)\bar{g}(\tau)y^{r-2}\,dx\,dy \right| < C \int_{y_0}^{\infty} dy \int_{\xi}^{\xi+\lambda} e^{-\alpha y}(a_0 + a_1 e^{-\beta y})y^{r-2}\,dx\,dy,$$

with certain positive constants α, β, and the integral in the right member converges. Next, replace S_i by a Stolz angle S_i'—a larger region—defined by $y/|x - p| > m > 0, y < y_1$. In S_i' we have, according to 3A, Exercise 1,

$$|g(u + iv)| < |u - p + iv|^{-r}\{b_0 + b_1 e^{-\gamma/v}\}, \qquad \gamma > 0$$

whereas f satisfies a similar estimate with $b_0 = 0$. Hence referring to Figure 25, we get, with $y = m|x - p|$,

$$\left| \iint_{S_i'} \right| < C \int_{-x_1}^{x_1} du \int_{y}^{y_1} v^{-r}e^{-\gamma/v} \cdot v^{-r} \cdot v^{r-2}\,dv < C \int_{-x_1}^{x_1} du \int_0^{y_1} v^{-r-2}e^{-\gamma/v}\,dv,$$

a convergent integral. Hence:

THEOREM. The integral (23) converges if both forms f, g belong to $C^+\{\Gamma, -r\}$ and at least one of them belongs to $C^0\{\Gamma, -r\}$.

COROLLARY. $C^0\{\Gamma, -r\}$ is a Hilbert space with the scalar product (23).

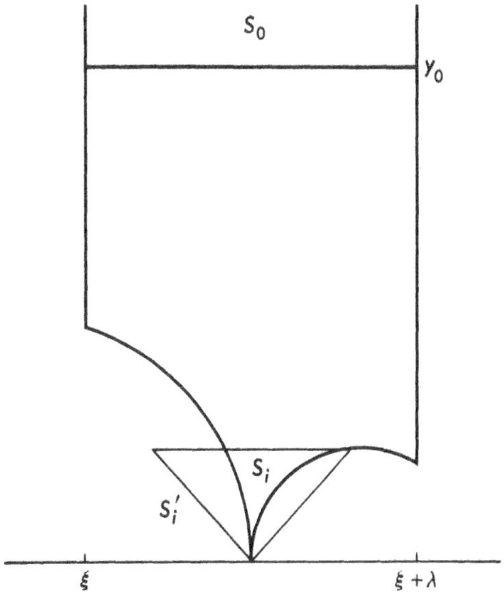

FIG. 25.

4C. The scalar product is independent of the choice of the normal polygon. Let $\tau = x + iy$, $\tau' = V\tau = x' + iy'$, where $V = (\cdot\,\cdot \mid c\; d) \in \Gamma$. Since $y' = y/|c\tau + d|^2$ and $dx'\,dy'/y'^2 = dx\,dy/y^2$ we have

$$f(\tau')\bar{g}(\tau')y'^r\,dx'dy'/y'^2 = (c\tau + d)^r f(\tau)\overline{(c\tau + d)}^r \bar{g}(\tau)y^r|\,c\tau + d\,|^{-2r}\,dxdy/y^2$$
$$= f(\tau)\bar{g}(\tau)y^r\,dxdy/y^2\,,$$

and so we have the

THEOREM. $(f, g;\, VN_0) = (f, g;\, N_0), \qquad V \in \Gamma.$

From now on we write (f, g) for the scalar product.

4D. Since we hope to establish the Poincaré series $\{G_{-r}(\tau, \nu)\}$ as a set of basis functions for $C^0\{\Gamma, -r\}$, it is essential that we calculate the scalar product of an arbitrary member of C^0 with G_{-r}. First, however, we shall derive two important estimates. For $f \in \{\Gamma, -r\}$ we note that

$$\varphi(\tau) = y^{r/2}\,|f(\tau)|, \qquad \tau = x + iy$$

is Γ-invariant, since with $V = (\cdot \cdot \mid c\ d) \in \Gamma$, $\tau' = V\tau = x' + iy'$, we have

$$\varphi(\tau') = y'^{r/2} \mid f(\tau') \mid = (y \mid c\tau + d \mid^{-2})^{r/2} \mid c\tau + d \mid^{r} \mid f(\tau) \mid = \varphi(\tau).$$

Now suppose f is a cusp form. Then f has positive order at each parabolic vertex p and 3A, Exercise 1 shows that $\varphi(\tau) \to 0$ as $\tau \to p$ from within a normal polygon N_0. Hence in a triangular region of N_0 with apex at p, φ is bounded, and in the remainder of N_0 (a relatively compact subset of H), φ is also bounded. Because Γ is an H-group there are only finitely many parabolic vertices in N_0. It follows that there is a constant C such that $\mid \varphi(\tau) \mid \leq C$ for τ in N_0. By invariance we deduce that $\mid \varphi(\tau) \mid \leq C$ in the whole upper half-plane; hence

$$\mid f(\tau) \mid \leq C y^{-r/2}, \qquad \tau \in \mathrm{H}.$$

This is the first estimate.

When f is expanded about $i\infty$,

$$f(\tau) = \sum_{h=\mu}^{\infty} a_h e\left(\frac{h\tau}{\lambda}\right), \tag{24}$$

the a_h are called the *Fourier coefficients* of f. By multiplying both members with $e(-m\tau/\lambda)$ and integrating, we find

$$\lambda a_m = \int_{\omega}^{\omega+\lambda} f(\tau) e\left(\frac{-m\tau}{\lambda}\right) d\tau,$$

where $\omega = u + iv$ may be any fixed point of H. Hence by the above estimate

$$\lambda \mid a_m \mid \leq \lambda C v^{-r/2} \exp\left(\frac{2\pi m v}{\lambda}\right).$$

In particular we can make v depend on m; choosing $v = 1/m$ we complete the proof of the following

THEOREM. For $f \in C^0\{\Gamma, -r\}$ we have

$$\mid f(\tau) \mid \leq C_1 y^{-r/2}, \qquad \tau \in \mathrm{H}$$
$$\mid a_m \mid \leq C_2 m^{r/2}, \qquad m = 1, 2, \ldots$$

where $\{a_m\}$ are the Fourier coefficients of f.

4E. The applications of the Hilbert space of cusp forms depend on the

SCALAR PRODUCT FORMULA. If $f \in C^0\{\Gamma, -r\}$, then

$$(f, G_{-r}(\tau, \nu)) = \begin{cases} a_\nu \nu^{1-r} e_r, & \nu > 0 \\ 0, & \nu = 0 \end{cases}$$

where a_ν is the νth Fourier coefficient of f [see (24)] and

$$e_r = 2\lambda^r(r-2)!(4\pi)^{1-r}.$$

We shall choose for N_0 an unbounded normal polygon. From formula (3) of 1A we get

$$(G,f) = \iint_{N_0} \sum_{M\in(\mathbf{D})} \frac{e(\nu M\tau/\lambda)}{(c\tau+d)^r} f(\tau) y^{r-2}\, dx\, dy, \qquad \mathbf{D} = \Gamma_\infty \setminus \Gamma.$$

We can select $(\mathbf{D})$ so that it contains both M and $-M$, and these matrices make the same contribution to the integral. Define $(\mathbf{D}')$ by the condition $c > 0$, or $c = 0$ and $a > 0$, in $M \in (\mathbf{D})$; then

$$(G,f) = 2\iint_{N_0} \sum_{M\in(\mathbf{D}')} \frac{e(\nu M\tau/\lambda)}{(c\tau+d)^r} f(\tau) y^{r-2}\, dx\, dy.$$

We wish to interchange sum and integral. Partition N_0 in the manner of 4B into a compact set S, a finite number of triangular sectors S_i at the finite parabolic vertices, and a portion of a vertical strip S_0. The interchange of order over S presents no difficulty since the series converges uniformly there (Theorem 1C). As in 4B we enlarge S_i to a Stolz angle S_i' bounded by $y/|x-p_i| > m > 0$, $y < y_1$. In S_i' we have $f(\tau) = f(u+iv) = O(e^{-\alpha/v})$, $\alpha > 0$, while $|c\tau+d|^{-r} = c^{-r}|\tau + d/c|^{-r} \le c^{-r}v^{-r}$ for $c > 0$; the single term with $c = 0$ may be disregarded. Moreover, the exponential in the numerator never exceeds 1 in absolute value. Hence the integral over S_i is majorized by

$$\int_{-x_1}^{x_1} du \int_{v}^{y_1} \left(\sum_M c^{-r}\right) e^{-\alpha/v} v^{-2}\, dv < C \int_0^{y_1} e^{-\alpha/v} v^{-2}\, dv < \infty,$$

since $\sum c^{-r}$ converges for $r > 2$. In the vertical strip S_0 we have $f(u+iv) = O(e^{-\beta v})$, $\beta > 0$, and so

$$\left|\iint_{S_0}\right| \le C \int_0^{\lambda} du \int_{y_0}^{\infty} \left(\sum_M c^{-r}\right) e^{-\beta v} v^{-2}\, dv < \infty.$$

Hence the interchange is legitimate and

$$(G,f) = 2 \sum_{M\in(\mathbf{D}')} \iint_{N_0} f(\tau) e\left(\frac{\nu M\tau}{\lambda}\right) (c\tau+d)^{-r} y^{r-2}\, dx\, dy.$$

In each integral in the right member we set $M\tau = M_i\tau = w = u + iv$; then with $N_i = M_i N_0$ we get

$$(G,f) = 2 \sum_{M_i\in(\mathbf{D}')} \iint_{N_i} f(w) e\left(\frac{\nu w}{\lambda}\right) v^{r-2}\, du\, dv, \tag{25}$$

since $f(\tau)(c\tau+d)^r = f(M\tau) = f(w)$.

We have chosen N_0 so that it lies in the strip B: $0 < x < \lambda$, $y > 0$. We now choose the matrices M_i of (D′) so that N_i lies in the strip B_1: $-\lambda < x < \lambda$, $y > 0$. If N_i lies in B we leave it alone. Otherwise write $N_i = N_i' \cup N_i''$, where $N_i' = N_i \cap B$. Now we observe that the integrand of (25) is invariant under the translation $w \to w + \lambda$; hence the integral over N_i'' is the same as the integral over $U^\lambda N_i''$, which lies in $\bar{B}$. Defining N_i^* to be N_i if $N_i \subset B$ and $N_i' \cup U^\lambda N_i''$ in the contrary case, we see that all N_i^* lie in $\bar{B}$ and in fact the union of the N_i^* covers B without overlapping except for their boundaries, a set of plane measure zero. We can therefore use the complete additivity of the Lebesgue integral to assert that

$$(G, f) = 2\int_0^\infty \int_0^\lambda \cdots = 2\lim_{\alpha\to 0}\int_\alpha^\infty \int_0^\lambda f(w) e\left(\frac{\nu w}{\lambda}\right) v^{r-2}\, du\, dv.$$

Insert the Fourier expansion (24) of f:

$$(G, f) = 2\lim_{\alpha\to 0}\int_\alpha^\infty \int_0^\lambda e\left(\frac{\nu w}{\lambda}\right) v^{r-2} \sum_{m=1}^\infty \bar{a}_m e\left(\frac{-m\bar{w}}{\lambda}\right) du\, dv. \tag{26}$$

We shall presently justify the inversion of sum and integral, and this gives

$$(G, f) = 2\lim_{\alpha\to 0}\sum_{m=1}^\infty \bar{a}_m \int_\alpha^\infty v^{r-2} \exp\left(\frac{-2\pi(\nu + m)v}{\lambda}\right) dv \int_0^\lambda e\left(\frac{(\nu - m)u}{\lambda}\right) du.$$

The inside integral vanishes unless $m = \nu$; hence for $\nu \neq 0$

$$\begin{aligned}(G, f) &= 2\lim_{\alpha\to 0}\lambda\bar{a}_\nu \int_\alpha^\infty v^{r-2}\exp\left(\frac{-4\pi\nu v}{\lambda}\right) dv \\ &= 2\lambda\bar{a}_\nu \int_0^\infty \cdots = 2\bar{a}_\nu \lambda^r (r-2)!(4\pi\nu)^{1-r}.\end{aligned}$$

If $\nu = 0$, the inside integral is always 0, since m is never equal to ν, and (G, f) vanishes. This is the desired result, since $(G, f) = \overline{(f, G)}$.

To complete the proof we must justify the order inversion in (26). The Fourier series $\sum \bar{a}_m e(-m\bar{w}/\lambda)$ converges uniformly in the region of integration, and it is therefore sufficient to show that

$$\int_\alpha^\infty \int_0^\lambda v^{r-2} \sum_m |a_m| \exp\left(\frac{-2\pi(\nu + m)v}{\lambda}\right) du\, dv$$

is finite. Now by Theorem 4D,

$$\begin{aligned}\sum_{m=1}^\infty |a_m| e^{-2\pi m v/\lambda} &< C\sum_1^\infty m^{r/2} e^{-2\pi m v/\lambda} \\ &\le C e^{-2\pi v/\lambda}\sum_1^\infty m^{r/2} e^{-2\pi(m-1)\alpha/\lambda} < C_1(\alpha) e^{-2\pi v/\lambda}\end{aligned}$$

for $v \geq \alpha > 0$. Hence

$$\int_{\alpha}^{\infty}\int_{0}^{\lambda}\sum_{m} < \lambda C_1(\alpha)\int_{\alpha}^{\infty} v^{r-2}e^{-2\pi v/\lambda}\,dv,$$

and this is finite for each fixed, positive α, as required.

4F. The remainder of this chapter will be devoted to applications of the Scalar Product Formula.

We recall an elementary result. If H is a Hilbert space of finite dimension, every linear manifold M in H is also finite dimensional and so is a subspace. Two elements $x, y \in H$ are called orthogonal if $(x, y) = 0$, and $(x, M) = 0$ means x is orthogonal to each element of M. Define

$$N = \{x \in H \mid (x, M) = 0\};$$

it is called the orthogonal complement of M and is easily proved to be a subspace. Moreover, H is the direct sum of M and N, by which we mean that every $x \in H$ has the unique decomposition $x = y + z, y \in M, z \in N$.

As a first application of the Scalar Product Formula we prove the

COMPLETENESS THEOREM. The space $C^0\{\Gamma, -r\}$ is spanned by the Poincaré series $\{G_{-r}(\tau, \nu), \nu > 0\}$.

A Poincaré series with $\nu > 0$ belongs to C^0 (see Theorem 1E). Since C^0 is finite dimensional (3E, Exercise 2), the linear manifold $\mathbf{G}$ generated by $\{G_{-r}(\tau, \nu), \nu > 0\}$ is a subspace of C^0. Let $\mathbf{H}$ be the orthogonal complement of $\mathbf{G}$ in C^0. Since $\mathbf{H} \subset C^0$, a form $f \in \mathbf{H}$ has Fourier coefficients $\{a_m, m \geq 1\}$. But $(f, \mathbf{G}) = 0$ and so by the Scalar Product Formula

$$0 = (f, G_{-r}(\tau, \nu)) = a_\nu \cdot \nu^{1-r}e_r, \qquad \nu = 1, 2, \cdots.$$

The Fourier coefficients of f all vanish and $f \equiv 0$. Hence $\mathbf{H} = \{0\}$, $\mathbf{G} = C^0$.

4G. Since the space of cusp forms is finite dimensional whereas there are infinitely many Poincaré series, there must be linear relations among them. Let G_{-r} be renormalized and then expanded in a Fourier series:

$$g(\tau, \nu) = \nu^{r-1}G_{-r}(\tau, \nu) = \sum_{m=1}^{\infty} b_m(\nu)e\left(\frac{m\tau}{\lambda}\right), \qquad \nu > 0; \tag{27}$$

the new functions $g(\tau, \nu)$ are still cusp forms. The Scalar Product Formula then reads

$$(f, g(\tau, \nu)) = a_\nu e_r,$$

where a_ν is, as usual, the νth Fourier coefficient of f.

THEOREM. The linear relation

$$\sum_{k=1}^{n} c_k g(\tau, \nu_k) \equiv 0 \tag{28}$$

holds if and only if

$$\sum_{i,k=1}^{n} c_i \bar{c}_k b_{\nu_k}(\nu_i) = 0.$$

Let $H(\tau)$ denote the left member of (28). Then $H \in C^0$ and

$$a_{\nu_k}(H) = \sum_{i=1}^{n} c_i b_{\nu_k}(\nu_i).$$

Hence

$$\begin{aligned}(H(\tau), H(\tau)) &= \sum_{k=1}^{n} \bar{c}_k (H(\tau), g(\tau, \nu_k)) \\ &= e_r \sum_k \bar{c}_k a_{\nu_k}(H) = e_r \sum_k \bar{c}_k \sum_i c_i b_{\nu_k}(\nu_i).\end{aligned}$$

The vanishing of the left member is equivalent to (28).

This gives immediately:

COROLLARY. The Poincaré series $g(\tau, \nu)$ vanishes identically if and only if

$$b_\nu(\nu) = 0.$$

We have merely to take $n = 1$, $\nu_1 = \nu$, $c_1 = 1$.

This is the first contribution to the problem of the identical vanishing of the Poincaré series.

Exercise 1. $b_\nu(\mu) = b_\mu(\nu)$.

4H. Let $\{\varphi_1(\tau), \cdots, \varphi_\mu(\tau)\}$ be a basis for C^0; that is, the forms φ_i span the space and are linearly independent. Let

$$\varphi_i(\tau) = \sum_{m=1}^{\infty} d_m(i) e\left(\frac{m\tau}{\lambda}\right), \qquad i = 1, \ldots, \mu$$

be the Fourier series of φ_i and define the μ-dimensional vectors

$$\mathbf{d}_m = (d_m(1), \cdots, d_m(\mu)).$$

THEOREM. The linear relation

$$\sum_{k=1}^{n} c_k g(\tau, \nu_k) \equiv 0 \tag{29}$$

holds if and only if

$$\sum_{k=1}^{n} \bar{c}_k \, \mathbf{d}_{\nu_k} = 0. \tag{30}$$

Denoting the left member of (29) by $H(\tau)$, we have

$$(\varphi_i(\tau), H(\tau)) = e_r \sum_{k=1}^{n} \bar{c}_k \, d_{\nu_k}(i), \qquad i = 1, 2, ..., \mu.$$

If $H(\tau) \equiv 0$, the right members all vanish, which implies $\sum_{k=1}^{n} \bar{c}_k \mathbf{d}_{\nu_k} = 0$. Conversely, if (30) is satisfied, H is orthogonal to each φ_i and so to all of C^0, since the $\{\varphi_i\}$ form a basis. Hence $H \equiv 0$.

4I. If

$$\mu = \dim C^0\{\Gamma, -r\},$$

there are μ Poincaré series $g(\tau, \nu_j), j = 1, 2, \cdots, \mu$, which form a basis for C^0, since we have seen in 4F that the Poincaré series with $\nu > 0$ span C^0. In general it is very difficult to determine the values ν_j in a basis, but for groups of genus zero (see I, 5D; III, 1I) we have the following result:

If Γ is a group of genus 0, *the first μ Poincaré series* $\{g(\tau, 1), \cdots, g(\tau, \mu)\}$ *form a basis for* $C^0\{\Gamma, -r\}$.

This striking result shows in particular that none of these Poincaré series vanishes identically, another contribution to our main problem. We shall give the proof only for the modular group, which in this section we denote by M.

Choose the usual fundamental region for M with cusp at $i\infty$ (see I, 6B). Suppose $\{g(\tau, i), i = 1, \cdots, \mu\}$ are linearly dependent; then by Theorem 4H there are $\{c_i\}$ not all zero such that

$$c_1 \mathbf{d}_1 + \cdots + c_\mu \mathbf{d}_\mu = 0,$$

the $\mathbf{d}_i$ being the Fourier vectors of a basis $\{\varphi_i\} \in C^0$. This is equivalent to

$$\sum_{m=1}^{\mu} c_m d_m^{(i)} = 0, \qquad i = 1, 2, ..., \mu$$

and by a theorem of algebra this implies

$$\sum_{m=1}^{\mu} c'_m d_i^{(m)} = 0, \qquad i = 1, 2, ..., \mu \tag{31}$$

with constants c'_m not all zero. The automorphic form $\Phi = c'_1\varphi_1 + \cdots + c'_\mu\varphi_\mu$ is therefore not identically zero and so it has finite order n_0 at $i\infty$. From (31) it is clear that

$$n_0 \geq \mu + 1. \tag{32}$$

We shall take certain results from III, 2K. The dimension of $C^+\{M, -2\} = 0$; this case is therefore disposed of and we assume $r > 2$. Now we assert

$$\dim C^0 = \dim C^+ - 1. \tag{33}$$

Indeed, $C^+ - C^0$ is not empty for it contains the Poincaré series $G_{-r}(\tau, 0)$—see Theorem 1E. Hence $\dim C^0 < \dim C^+$. On the other hand suppose $f_1, f_2, \cdots, f_\mu$ is a basis for C^+ and let $f_1, \cdots, f_\nu$ be the nonidentically vanishing cusp forms among them with $\nu \leq \mu - 2$. Replace the f's with the set

$$f_1, f_2, \cdots f_{\mu-2}, \alpha f_{\mu-1} + \beta f_\mu, f_\mu,$$

where $\alpha \neq 0, \beta \neq 0$ are chosen so that $\alpha f_{\mu-1} + \beta f_\mu$ vanishes at $i\infty$. It is apparent that the new set of functions is also linearly independent, but it contains one more cusp form than the set $f_1, \cdots, f_\nu$. This contradiction proves the assertion (33).

Reference to III, 2K now shows that

$$\mu = \begin{cases} \left[\frac{r}{12}\right] - 1, & r \equiv 2 \pmod{12} \\ \left[\frac{r}{12}\right], & r \not\equiv 2 \pmod{12} \end{cases} \tag{34}$$

But by II, Theorem 3D, $\deg \Phi = r/12$, for there are two elliptic cycles, of orders 2 and 3, and one parabolic cycle in a fundamental region of M (see I, 6B) and the genus of M is 0 (see III, 2K). Since Φ has no poles, the order n_0 of Φ at $i\infty$ satisfies $n_0 \leq [r/12]$, for n_0 is integral.

Now when $r \not\equiv 2 \pmod{12}$ we have

$$n_0 \leq \left[\frac{r}{12}\right] = \mu < \mu + 1,$$

a contradiction to (32).

Finally, we must treat the case $r \equiv 2 \pmod{12}$. At the elliptic fixed points of orders 2 and 3 the order of Φ is $k/2$ and $l/3$, respectively, k and l being integers (Theorem 3A). Suppose $n_0 = [r/12]$; then from $\deg \Phi = r/12$ we find

$$\frac{k}{2} + \frac{l}{3} = \frac{r}{12} - n_0 = \frac{r}{12} - \left[\frac{r}{12}\right] = \frac{1}{6}.$$

This equation cannot be solved in nonnegative integers k, l. Hence $n_0 < [r/12]$, and (34) gives $n_0 < \mu + 1$, again a contradiction.

We have proved:

THEOREM. If M is the modular group, $r \geq 2$ is an even integer, and μ is the dimension of the space $C^0\{M, -r\}$, the first μ Poincaré series $g(\tau, 1), \cdots, g(\tau, \mu)$ form a basis for $C^0\{M, -r\}$.

4J. Another criterion for the identical vanishing of the Poincaré series is contained in the following

THEOREM. The Poincaré series $g(\tau, \nu)$ of dimension $-r$ vanishes identically if and only if every cusp form of dimension $-r$ has a zero νth Fourier coefficient.

Let the Fourier coefficients of $f \in C^0\{\Gamma, -r\}$ be $a_1(f), a_2(f), \cdots$. If $g(\tau, \nu) \equiv 0$, then

$$e_r a_\nu(f) = (f, g(\tau, \nu)) = 0$$

for each cusp form f. Conversely suppose $a_\nu(f) = 0$ for all f. In particular this holds for a basis $(\varphi_1, \cdots, \varphi_\mu)$ of C^0. Then in Theorem 4H we have $\mathbf{d}_\nu = 0$, which implies $g(\tau, \nu) \equiv 0$.

4K. We are now able to characterize the Poincaré series within the space of cusp forms.

THEOREM. Up to a multiplicative constant, the Poincaré series $g(\tau, \nu)$ is precisely that cusp form which is orthogonal to the linear manifold of cusp forms having vanishing νth Fourier coefficients.

Let

$$\mathbf{D}_\nu = \{f \in C^0 \mid a_\nu(f) = 0\}$$

be the manifold of cusp forms referred to in the theorem and let $\mathbf{N}_\nu$ be the orthogonal complement of $\mathbf{D}_\nu$ in C^0. The theorem asserts that $h \in \mathbf{N}_\nu$ implies $h = cg(\tau, \nu)$, $c =$ constant. For abbreviation write $g_\nu = g(\tau, \nu)$.

Suppose $g_\nu \equiv 0$. By Theorem 4J every $f \in C^0$ has $a_\nu(f) = 0$. Then $\mathbf{D}_\nu = C^0$ and $\mathbf{N}_\nu = \{0\}$: the result is trivial.

Hence assume $g_\nu \not\equiv 0$ and define

$$k = h - \frac{(h, g_\nu)}{(g_\nu, g_\nu)} g_\nu, \qquad h \in \mathbf{N}_\nu. \tag{35}$$

Then

$$e_r a_\nu(k) = (k, g_\nu) = (h, g_\nu) - (h, g_\nu) = 0. \tag{36}$$

It follows that $k \in \mathbf{D}_\nu$. Hence $(k, h) = 0$, for $h \in \mathbf{N}_\nu$. From (35) we now get

$$(k, k) = (h, k) - \frac{(h, g_\nu)}{(g_\nu, g_\nu)} (g_\nu, k) = 0,$$

since $(k, g_\nu) = 0$ by (36). But $k = 0$ together with (35) implies the desired result.

4L. One of the most beautiful applications of the scalar product is to the Hecke-Petersson theory of Dirichlet series with Euler products. This theory is too extensive to be given here but we include a brief resumé in the note to this chapter (page 108).

Note to Chapter 2.

The Riemann zeta function $\zeta(s)$ has the following properties:

(1) It can be expanded in a convergent Dirichlet series $\sum_{n=1}^{\infty} n^{-s}$.

(2) It has a continuation to the whole plane and $(s-1)\zeta(s)$ is an entire function of finite order.

(3) It satisfies a functional equation:

$$R(s) = \pi^{-s}\Gamma(s)\zeta(2s) = R(\tfrac{1}{2} - s).$$

But conversely, these properties determine $\zeta(s)$ uniquely up to a constant factor, as was shown by H. Hamburger (1921).

E. Hecke proposed the following generalization. Let three numbers be given:

$$\{\lambda, k, \gamma\}$$

with $\lambda > 0$, $k > 0$. Find all functions $\varphi(s)$ such that:

(1′) $\varphi(s)$ can be expanded in a Dirichlet series that converges somewhere.

(2′) $\varphi(s)$ has a continuation to the whole plane and $(s-k)\varphi(s)$ is an entire function of finite order.

(3′) $\varphi(s)$ satisfies a functional equation

$$R(s) = \left(\frac{2\pi}{\lambda}\right)^{-s} \Gamma(s)\varphi(s) = \gamma R(k-s).$$

By iteration of this equation we see at once that $\gamma = \pm 1$.

The solution to this problem leads to functions and conclusions of number-theoretic interest. Hecke has found all "signatures" $\{\lambda, k, \gamma\}$ that lead to nontrivial solutions (Math. Ann. vols. 112, 114). The zeta function $\zeta(s)$ has signature $\{2, \frac{1}{2}, 1\}$. We shall consider only the case

$$\{1, k, (-1)^{k/2}\}, \qquad k \text{ even}, \geq 4.$$

In the investigation the theory of modular forms† is used as a tool. But thereby is revealed an intimate connection between Dirichlet series and modular forms, a connection already observed in special cases by Riemann and Ramanujan.

We shall find that there are κ linearly independent solutions $\varphi(s)$ to the problem $\{1, k, (-1)^{k/2}\}$, where κ is the dimension of $C^+\{\Gamma, -k\}$ and Γ is the modular group. The solutions span a vector space Φ. Thus $\varphi(s)$ is not uniquely determined: any element of Φ is a solution.

---

† A modular form is an automorphic form on the modular group Γ.

For the unique determination of $\varphi(s)$ we need one other property possessed by $\zeta(s)$; namely, $\varphi(s)$ can be expanded in an Euler product:

$$\varphi(s) = \sum_{n=1}^{\infty} c_n n^{-s} = \prod_p (1 - c_p p^{-s} + p^{k-1-2s})^{-1}, \tag{4}$$

the product being extended over all primes p. There are exactly κ functions in Φ that have property (4), as we shall see.

Dirichlet Series and Modular Forms

The basic idea is to associate to each Dirichlet series

$$\varphi(s) = \sum_{n=1}^{\infty} c_n n^{-s}, \qquad s = \sigma + it$$

the Fourier series

$$F(\tau) = c_0 + \sum_{n=1}^{\infty} c_n e^{2\pi i n \tau}$$

(where c_0 is still to be determined), and conversely. The association can be made analytically by a "functional transformation":

$$(2\pi)^{-s}\Gamma(s)\varphi(s) \equiv R(s) = \int_0^{\infty} (F(iy) - c_0)y^{s-1}\, dy, \tag{$\dagger$}$$

and conversely,

$$F(y) = \frac{1}{2\pi i}\int_{\sigma_1 - i\infty}^{\sigma_1 + i\infty} R(s) y^{-s}\, ds, \qquad \sigma_1 > k - 1. \tag{$\ddagger$}$$

These relations correspond to the "Mellin transform pair":

$$\Gamma(s) = \int_0^{\infty} e^{-y} y^{s-1}\, dy,$$

$$e^{-y} = \frac{1}{2\pi i}\int_{\sigma_1 - i\infty}^{\sigma_1 + i\infty} \Gamma(s) y^{-s}\, ds.$$

Obviously, linearly independent Dirichlet series are associated to linearly independent functions $F(\tau)$.

If φ satisfies $(1') - (3')$, then $F(\tau)$ is a modular form: $F \in C^{+}\{\Gamma, -k\}$. The functional equation $(3')$ corresponds to the transformation equation of F. The proof requires an estimate: $\varphi(\sigma + it) = O(|\,t\,|^{A})$, $|\,t\,| > 1$, uniformly in any strip $\beta_1 \leq \sigma \leq \beta_2$, which is a consequence of $(1') - (3')$ and the Phragmén-Lindelöf principle. The coefficient c_0, undefined by φ, appears naturally in the course of the proof.

Conversely, if $F \in C^+\{\Gamma, -k\}$, the Fourier coefficients c_n have the estimate $c_n = O(n^{k-1})$, as shown in 4D; and this enables us to prove that $\varphi(s)$ satisfies (1′) – (3′).

We can now say that the solutions $\varphi(s)$ of our problem span a vector space Φ of dimension $\kappa = \dim C^+\{\Gamma, -k\}$. Moreover, Φ and $C^+\{\Gamma, -k\}$ are isomorphic through the functional transformation (†), (‡).

Euler Products

We shall now impose the requirement that the solution $\varphi(s) = \sum c(n)n^{-s}$ have an Euler product (4). As a consequence we can show first that $c(n)$ is multiplicative:

$$c(mn) = c(m)c(n) \qquad \text{for } (m, n) = 1, \tag{5}$$

in particular,

$$c(1) = 1. \tag{5'}$$

Next we have for each prime p and for $m \geq 1$,

$$c(p)c(p^m) = c(p^{m+1}) + p^{k-1}c(p^{m-1}). \tag{6}$$

Conversely, (5) and (6) are sufficient to insure that the absolutely convergent series $\sum c(n)n^{-s}$ has an Euler product.

From (5) and (6) we deduce by induction on the number of distinct primes dividing both m and n that

$$c(m)c(n) = \sum_{d\mid(m,n)} d^{k-1}c\left(\frac{mn}{d^2}\right), \tag{7}$$

and, conversely, (7) implies (5) and (6). Because of the correspondence between Dirichlet series and modular forms, it is sufficient to find forms whose Fourier coefficients satisfy (7). Moreover, if there are Dirichlet series with Euler products, such modular forms necessarily exist.

Hecke's Operators

For this purpose Hecke introduced the T_n operators. The set of matrices $(a\, b \mid c\, d)$ with integral entries and determinant $ad - bc = n \geq 1$ fall into a finite number of equivalence classes with respect to left multiplication by elements of Γ. As representatives of these classes we may take the set

$$M = \left\{ \begin{pmatrix} a & b \\ 0 & d \end{pmatrix} \,\middle|\, ad = n, \quad d > 0, \quad b \bmod d \right\}.$$

Then we define, for $F \in C^+\{\Gamma, -k\}$,

$$F \mid T_n = n^{k-1} \sum_M d^{-k} F\left(\frac{a\tau + b}{d}\right).$$

The $\{T_n\}$ form a ring of operators with identity T_1 having the following properties:

(i) Each T_n maps C^+ into C^+ and maps C^0 into C^0.

(ii) All T_n are commutative.

(iii) $T(n)T(m) = \sum_{d \mid (m,n)} d^{k-1} T\left(\frac{mn}{d^2}\right);$

in particular

$$T(n)T(m) = T(mn) \qquad \text{for } (m, n) = 1.$$

(iv) If $F(\tau) = \sum_{m=0}^{\infty} c(m) e^{2\pi i m\tau}$, we have

$$F \mid T_n = \sum_{m=0}^{\infty} e^{2\pi i m\tau} \sum_{d \mid (m,n)} c\left(\frac{mn}{d^2}\right) d^{k-1}.$$

The last equation gives the clue. We recognize that *the coefficients of* $F(\tau)$ *satisfy* (7) *if and only if* F *is an eigenfunction of all the operators* $\{T_n\}$. Indeed, (7) and (iv) imply

$$F \mid T_n = \sum_{m=0}^{\infty} e^{2\pi i m\tau} \sum_{d \mid (m,n)} c\left(\frac{mn}{d^2}\right) d^{k-1} = c(n) \sum_{m=0}^{\infty} c(m) e^{2\pi i m\tau} = c(n)F.$$

Conversely, if F is an eigenfunction of T_n with eigenvalue $\rho(n)$, we get

$$\sum_{d \mid (m,n)} c\left(\frac{mn}{d^2}\right) d^{k-1} = \rho(n)c(m),$$

and for $m = 1$ this yields $c(n) = \rho(n)c(1) = \rho(n)$, in other words, (7).

We have now reduced our problem to one of functional analysis: *find the eigenfunctions of the ring of operators* $\{T_n\}$ *acting on the vector space* $C^+\{\Gamma, -k\}$.

Petersson's Hilbert Space of Cusp Forms

Let us observe at once that we can restrict our considerations from the beginning to $C^0\{\Gamma, -k\}$. For we know an eigenfunction lying in $C^+ - C^0$; it is the Eisenstein series $G_{-k}(\tau, 0)$, which we now normalize as (2C, Exercise 2)

$$g(\tau, 0) = c_0 + \sum_{m=1}^{\infty} \sigma_{k-1}(m) e^{2\pi i m\tau}, \qquad \sigma_{k-1}(m) = \sum_{d \mid m} d^{k-1}.$$

Here $\{c(n) = \sigma_{k-1}(n), n > 1\}$ has the properties (5), (6), as is trivially verified, and therefore satisfies (7). It follows that $g(\tau, 0)$ is an eigenfunction of the ring $\{T_n\}$ and so we have the Euler product

$$\varphi(s) = \sum_1^\infty \frac{\sigma_{k-1}(n)}{n^s} = \zeta(s)\zeta(s-k+1) = \prod_p (1 - \sigma_{k-1}(p)p^{-s} + p^{k-1-2s}).$$

We can now concentrate on C^0. The dimension of C^0 is $\mu = \kappa - 1$, as we saw in 4I, (33).

At this point the application of the *Hilbert* space C^0 enters the picture. Petersson studied the action of the operator T_n on C^0 and proved

$$(f \mid T_n, g) = (f, g \mid T_n), f, g \in C^0. \tag{8}$$

This says that T_n is a hermitian operator on C^0. It is the crucial result that we need, for it enables us to simultaneously diagonalize the whole ring $\{T_n\}$.

In the proof of (8) we may assume f, g are two Poincaré series $g_\nu(\tau) = g(\tau, \nu)$, $\nu > 0$, since these functions span C^0 (see 4F, (27)). Now we can prove in turn:

$$c_\mu(\nu, n) = \sum_{d \mid (n,\mu)} d^{k-1} c_{\mu n/d^2}(\nu) = c_n(\nu, \mu),$$

where $c_\mu(\nu, n)$ is the μth Fourier coefficient of $g_\nu \mid T_m$;

$$g_\nu \mid T_n = \sum_{d \mid (n,\nu)} d^{k-1} g\left(\tau, \frac{n\nu}{d^2}\right) = g_n \mid T_\nu,$$

from which follows

$$c_m(\nu, n) = \sum_{d \mid (n,\nu)} d^{k-1} c_m\left(\frac{n\nu}{d^2}\right) = c_m(n, \nu).$$

Then

$$\begin{aligned}(g_\nu \mid T_n, g_\mu) &= e_k c_\mu(\nu, n) = e_k c_n(\nu, \mu) = e_k c_n(\mu, \nu)\\ &= e_k c_\nu(\mu, n) = (g_\mu \mid T_n, g_\nu)\\ &= \text{conjugate of}\,(g_\nu, g_\mu \mid T_n) = (g_\nu, g_\mu \mid T_n),\end{aligned}$$

since it turns out that $(g_\nu, g_\mu \mid T_n)$ is real. This proves (8).

Conclusion

We now proceed as follows. Let $F_1(\tau), F_2(\tau), \cdots, F_\mu(\tau)$ be an orthonormal basis for C^0. Write

$$F_\rho \mid T_n = \sum_{\sigma=1}^{\mu} \lambda_{\rho\sigma} F_\sigma, \qquad \rho = 1, 2, \ldots, \mu.$$

Define the $\mu \times \mu$ matrix

$$\lambda(n) = (\lambda_{\rho\sigma}(n)), \qquad n = 1, 2, \cdots.$$

Then the mapping $T(n) \to \lambda(n)$ is a faithful representation of $\{T(n)\}$. The properties of $\{T(n)\}$ carry over to $\{\lambda(n)\}$: all $\lambda(n)$ are commutative and hermitian:

$$\lambda_{\rho\sigma}(n) = \bar{\lambda}_{\sigma\rho}(n), \qquad \sigma, \rho = 1, \cdots, \mu; \qquad n = 1, 2, \cdots. \tag{9}$$

Since $\lambda(n)$ is of order μ, there are at most μ^2 linearly independent matrices in $\{\lambda(n)\}$. By a theorem of algebra, there exists a $\mu \times \mu$ unitary matrix $A = (a_{\rho\sigma})$ such that

$$A\lambda(n)A^{-1} = D_n = \text{diagonal matrix},$$

first for the linearly independent matrices and therefore for all matrices in $\{\lambda(n)\}$. Setting

$$D_n = (\Lambda_\rho(n)\delta_{\rho\sigma}), \qquad n = 1, 2, \cdots$$

$$H_\rho(\tau) = \sum_{\sigma=1}^{\mu} a_{\rho\sigma}F_\sigma(\tau), \qquad \rho = 1, \ldots, \mu$$

we have the desired eigenfunctions:

$$H_\rho(\tau) \mid T_n = \Lambda_\rho(n)H_\rho(\tau), \qquad \rho = 1, 2, \cdots, \mu.$$

The $\{H_\rho\}$ are also an orthonormal basis for C^0, since A is unitary.

As a consequence of the hermitian character of T_n, $\Lambda_\rho(n)$ is real. We may assume the first Fourier coefficient of H_ρ is not zero, otherwise we could prove $H_\rho \equiv 0$. Then we can normalize H_ρ so that its first coefficient is 1. Now using the fact that H_ρ is an eigenfunction, we readily find that its Fourier coefficients are exactly $\Lambda_\rho(n)$:

$$K_\rho(\tau) = \sum_{n=1}^{\infty} \Lambda_\rho(n)e^{2\pi i n\tau}, \qquad \rho = 1, \ldots, \mu$$

where K_ρ is the normalized H_ρ. Since K_ρ is an eigenfunction, the Fourier coefficients $\Lambda_\rho(n)$ are multiplicative in the sense of (7).

Finally, it can be shown that the $\{K_\rho\}$ are determined uniquely apart from order.

We now summarize in a

MAIN THEOREM. *Let $k \geq 4$ be an even integer. The functions $\varphi(s)$ satisfying (1′) — (3′) span a vector space Φ of dimension κ, and Φ is isomorphic to the space $C^+\{\Gamma, -k\}$. In the space Φ there are exactly κ functions that have an Euler product (4); their coefficients $c_\rho(n)$, $\rho = 1, \cdots, \kappa$, are multiplicative in the sense of (7). One of these functions is $\zeta(s)\zeta(s - k + 1)$; the others are entire functions.*

The above theorem has been stated in terms of Dirichlet series, since our original problem involved only Dirichlet series. But our investigation sheds light on the space of modular forms, and we state these results in a final

THEOREM. In $C^+\{\Gamma, -k\}$ there are exactly κ modular forms $K_\rho(\tau)$ whose Fourier coefficients $\Lambda_\rho(n)$, $n = 1, 2, \cdots$, $\rho = 1, 2, \cdots, \kappa$, are, for fixed ρ, multiplicative functions of n in the sense of (7). The K_ρ are simultaneous eigenfunctions of the whole ring of Hecke operators $\{T_n\}$ and the eigenvalues are the Fourier coefficients $\Lambda_\rho(n)$. The K_ρ are uniquely determined apart from order, and they form a basis for C^+. There are $\kappa - 1$ of the functions K_ρ that form an orthogonal basis for C^0.

[III]

Riemann Surfaces

In this chapter we shall explore the connection between automorphic function theory and the theory of Riemann surfaces. Briefly it is this: given a real discrete group Γ, we identify the points of H^+ that are equivalent under Γ. The resulting set $S = \Gamma\backslash H^+$ can be endowed with a topological and analytic structure and thus becomes a Riemann surface. There is a projection mapping $\sigma: H^+ \rightarrow S$ that sends all points of an orbit Γx of Γ into a single point of S. Then an analytic function φ on S can be lifted to an automorphic function f on H^+ by the definition

$$f(\tau) = \varphi(\sigma\tau),$$

and, conversely, an automorphic function on H^+ projects into an analytic function on S. Likewise, differentials on S correspond to automorphic forms of dimension -2 on H^+, with the differentials of the first kind going into cusp forms. By this correspondence the theory of Riemann surfaces can be applied to automorphic function theory, and conversely.

It will be necessary to quote some results from Riemann surface theory whose proofs are too long to be included. Reference will be made to the textbooks of Ahlfors-Sario and Springer listed in the References at the end of this book.

1. The Quotient Space of H by a Group

1A. Let Γ be a real discrete group, horocyclic or not. Let P be the set of its parabolic vertices, and let $H^+ = H \cup P$. We shall construct a topology for H^+ by defining the sets S_x of a basis.[1]

For $x \in H$ let S_x be an open disk lying in H and containing x. For $x \in P$, x finite, define K to be a horocycle at x (a euclidean circle tangent to E at x), and set (see Figure 26)

$$S_x = \text{Int}\, K \cup \{x\}.$$

For $x = \infty$ define

$$S_\infty = \{z \mid \text{Im}\, z > h > 0\} \cup \{\infty\}.$$

Then S_x is defined for all $x \in H^+$. A set is defined to be open if and only if it is a union of basis sets S_x. It is not hard to verify that this yields a valid topology. Clearly it is a Hausdorff topology; that is, different points of H^+ can be covered by nonoverlapping open sets.

H^+ *is a connected space.* A space S is connected if it cannot be written as

$$S = A \cup B,$$

where A and B are open, disjoint, nonempty subsets of S. Now every point of H^+ can be joined to the point $z = i$ by a straight line that lies entirely in H^+. Each straight line is a connected set and it is well known that the union of connected sets all containing a given point is itself connected.

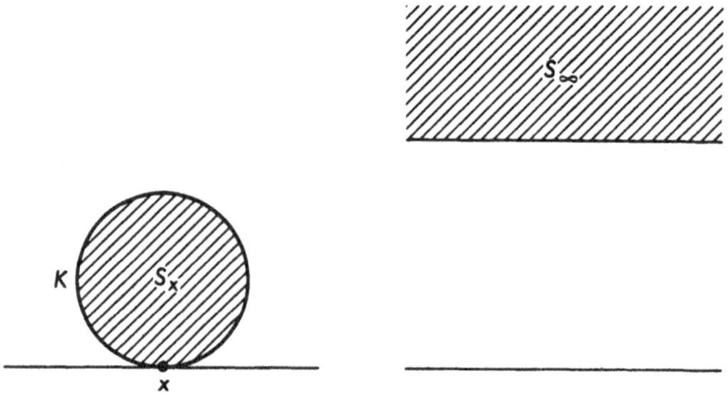

FIG. 26.

H^+ *is not locally compact.* Indeed, no $p \in P$ has a compact neighborhood. Suppose N is a compact neighborhood of p, then N must contain an open set containing p and so N contains an S_p. Since H^+ is Hausdorff, $\bar{S}_p$ is compact; therefore every infinite subset of $\bar{S}_p$ has a limit point in $\bar{S}_p$. Choose for the infinite subset a sequence of distinct points on the rim of S_p converging in the euclidean sense to p; call this sequence $\{z_n\}$. Then if $\{z_n\}$ has a limit point, it must be p. We can find another S'_p lying entirely within S_p; hence S'_p will contain no z_n. Since S'_p is a neighborhood of p, we have a contradiction.

It must be kept in mind that despite appearances p is an *interior* point of S_p.

Finally, note that the elements of Γ map open sets on open sets. We need verify this only for basis sets. But if $V \in \Gamma$ and $x \in H^+$, VS_x is a set of the form S_{Vx}.

1B. The Riemann surface is defined abstractly. It arises in analytic function theory through the process of continuation of a function element. An analytic function is uniquely defined by any one of its function elements. A function element, in turn, is associated with a certain domain. When the function is continued analytically, these domains determine the carrier of the function, and this carrier turns out to be a Riemann surface. See Springer, Chapter 3, where it is shown that every analytic function is associated with a Riemann surface on which it is single-valued.

We now make the formal

DEFINITION. A Riemann surface is a surface with directly conformal neighbor relations.

This means the following. Let S be a connected Hausdorff space. Let S be covered by a collection of open sets $\{U_\alpha\}$, and to each U_α let there be given a mapping Φ_α, which is a homeomorphism of U_α onto an open subset of the euclidean plane. This makes S a *surface*. A *neighbor relation* is a function

$$= \Phi_\beta \circ \Phi_\alpha^{-1}.$$

f is defined when U_α and U_β overlap and is a mapping from $\Phi_\alpha(U_\alpha \cap U_\beta)$ onto $\Phi_\beta(U_\alpha \cap U_\beta)$; that is, f is a mapping from one complex plane to another. It is one-to-one since it is the composition of homeomorphisms. We demand that each neighbor relation, when defined, be directly conformal on its domain; that is, $\Phi_\beta \circ \Phi_\alpha^{-1}$ must be analytic† on its domain.

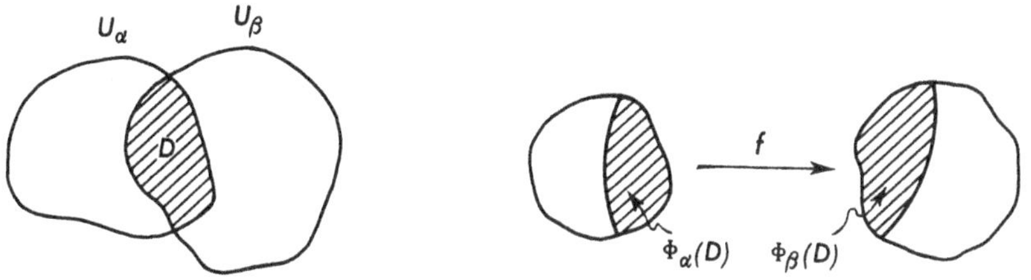

FIG. 27.

The sets U_α are called parametric neighborhoods, and every function

$$t_\alpha = \Phi_\alpha(q), \qquad q \in U_\alpha$$

is called a local variable at q. The homeomorphism Φ_α assigns a coordinate system to U_α. A point lying in two neighborhoods U_α, U_β will be covered by two coordinate systems, and the condition on the neighbor relations in the definition insures that the transition from one coordinate system to another is made in an analytic manner.

1C. We next construct the orbit space $\Gamma\backslash H^+$. Let Γx, the orbit of x, be denoted by $[x]$. Let

$$S = \{[x],\ x \in H^+\}.$$

† The condition that the derivative of $\Phi_\beta \circ \Phi_\alpha^{-1}$ be different from zero need not be assumed explicitly, since the function is known to be one-to-one.

That is, the points of S are the orbits in H^+; S is the set obtained by identifying points in H^+ that are equivalent under Γ. It is clear that S is in one-to-one correspondence with a fundamental set for Γ relative to H^+, and so we may take as a model of S a normal polygon to which certain sides and vertices have been added.

If X is an arbitrary topological space and f a function from X into the *set* Y, we can make Y a topological space in a natural way by defining a set A to be open in Y if and only if the inverse image $f^{-1}(A)$ is open in X. It is easy to check that this gives a valid topology for Y; it is called the topology induced by f. In this topology f is continuous.

Let us now define a mapping from H^+ to S by setting

$$\sigma(\tau) = [\tau], \tau \in H^+.$$

We see that σ is onto. If we now put on S the topology induced by σ, then σ is a continuous function on H^+. From now on when we speak of S we shall always mean S with the topology induced by σ.

Now the image of a connected space by a continuous mapping is connected; *hence S is a connected space.*

1D. We now prove that σ is an open mapping; that is, σ sends open sets into open sets. Let $A \subset H^+$ be open. Now $\sigma^{-1}\{\sigma(A)\} = \Gamma A$. Since the right member is open (for VA is open), so is the left. From the definition of open sets in S, we conclude that $\sigma(A)$ is open.

1E. If X is a Hausdorff space and f a mapping from $X \to Y$, it is not always the case that Y with the induced topology is Hausdorff. Therefore we must verify in our case that S is a Hausdorff space. Let $[x]$ and $[y]$ be distinct points of S; then $x \notin \Gamma y$.

Let N_0 be any normal polygon of Γ. We may assume x and y lie in $\bar{N}_0$, since $[Vx] = [x]$ for $V \in \Gamma$. There are several cases.

(1) $x, y \in \text{Int } N_0$. Obviously there exist closed disks $\bar{S}_x$, $\bar{S}_y$ that are disjoint and lie in N_0. Writing $X = \bar{S}_x$, $Y = \bar{S}_y$, we can assert that X, Y are compact neighborhoods of x, y, respectively, and

$$\Gamma X \cap \Gamma Y = \emptyset.$$

(2) $x \in \text{Int } N_0, y \in \text{Bd } N_0$ and is an inner point of a side. Let $y' \neq y$ be the (unique) point of $\bar{N}_0$ equivalent to y. Let $X \subset N_0$, Y, Y' be compact neighborhoods of x, y, y', respectively, which do not overlap. Then ΓX is disjoint from ΓY.

(3) x and y are inner points of sides. Here we use the same argument as in (2).

(4) x is an ordinary vertex of N_0, $y \in \text{Int } N_0$. Suppose x determines the (ordinary) cycle $\{x_1, \cdots, x_s\}$. Draw closed disks about x_i, $i = 1, \cdots, s$, equivalent

under Γ, and a closed disk Y about y, such that none of the disks overlap. If X is the disk about x, X and Y are compact and $\Gamma X \cap \Gamma Y = \phi$.

(5) x as in (4), $y \in \mathrm{H} \cap \bar{N}_0$. Use the argument of (4).

(6) $x, y \in \mathrm{P}$. This is the most troublesome case. Let $\{x_1, \cdots, x_s\}, \{y_1, \cdots, y_t\}$ be the cycles determined by x and y, respectively. A set S_x (see 1A) cuts out an open curvilinear triangle Δ_1 lying in N_0 and having the vertex x; Δ_1 has triangular images $\Delta_2, \cdots, \Delta_s$ at $x_2, \cdots, x_s$, respectively. Similarly S_y cuts out a triangle Δ_1' at y_1 with images $\Delta_2', \cdots, \Delta_t'$ at $y_2, \cdots, y_t$. S_x consists entirely of images of $\{\Delta_i\}$; S_y, entirely of images of $\{\Delta_j'\}$, as we saw in I, 4I. We select S_x, S_y so small that the sets $\{\Delta_i\}, \{\Delta_j'\}$ are mutually disjoint.

We assert ΓS_y does not meet ΓS_x. If not, $V\Delta_k$ meets $W\Delta_l'$ for some $V, W \in \Gamma$ and some k, l. Since $V\Delta_k$ lies entirely in one normal polygon and likewise $W\Delta_l'$, they must lie in the *same* normal polygon. Hence $V = W$ and Δ_k intersects Δ_l', a contradiction.

(7) $x \in \mathrm{P}$, $y \in \bar{N}_0$. Use the preceding arguments.

In all cases then, there exist compact neighborhoods X of x and Y of y such that $\Gamma X \cap \Gamma Y = \emptyset$. Hence $\sigma(X) \cap \sigma(Y) = \emptyset$. But $\sigma(X)$ is a neighborhood of $[x]$. For X contains an *open* neighborhood A of x, so

$$[x] \in \sigma(A) \subset \sigma(X),$$

and $\sigma(A)$ is open since σ is an open mapping. Similarly $\sigma(Y)$ is a neighborhood of $[y]$. This concludes the proof that S is a Hausdorff space.

1F. We are ready to introduce a two-dimensional structure on S: for each $[x] \in S$ we must exhibit an open neighborhood U_x of $[x]$ and a homeomorphism Φ_x that maps U_x onto an open set in the euclidean plane. For U_x we take simply

$$U_x = \sigma(S_x^*),$$

where S_x^* is a specially selected set in the basis for the topology of H^+. Since S_x^* is open, U_x is an open neighborhood of $[x]$.

We demand of S_x^* that $S_x^* - x$ contain no fixed point of Γ and that S_x^* contain no distinct Γ-equivalent points except possibly points equivalent by elements of Γ that fix x. Besides this we require that S_x^* be relatively compact if $x \in \mathrm{H}$. It is easy to see that we can meet these requirements. Moreover, if x is a fixed point, we choose S_x^* so that its boundary is a fixed circle of an element of Γ fixing x.

Now define the mappings†

$$\sigma_x = \text{restriction of } \sigma \text{ to } S_x^*; \qquad \Phi_x = \tau_x \circ \sigma_x^{-1}.$$

† If B is any set, $\sigma_x^{-1}B$ means $\sigma^{-1}B \cap S_x^*$.

Here $\tau_x = \tau_x(z)$ is from S_x^* to the plane; it will be defined presently. Φ_x will be the desired homeomorphism.

We can say at once that σ_x is continuous, for it is the restriction of a continuous function. We prove it is open. Let A be open in S_x^*; it is absolutely open because S_x^* is open. Then $\sigma_x A = \sigma A$ is open in S, since σ is an open mapping. For all $x \in \mathrm{H}^+$, σ_x is an open, continuous mapping of S_x^* into S. *Hence, in order to show Φ_x is a homeomorphism, it is sufficient to prove that it is one-to-one and that τ_x is open and continuous.*

(1) $x =$ nonfixed point; define $\tau_x(z) = z - x$. Then $\Phi_x(U_x) = \tau_x(S_x^*)$ is a disk in the z-plane. Trivially τ_x is open and continuous, also one-to-one. Also σ_x is one-to-one, since we have demanded that S_x^* contain no distinct equivalent points. Hence Φ_x is one-to-one.

(2) $x =$ elliptic fixed point of order l; define

$$\tau_x(z) = \left(\frac{z - x}{z - \bar{x}}\right)^l.$$

Here τ_x maps S_x^* in an l-to-1 manner on an open disk D and τ_x^{-1} carries a point w of D into l equivalent points $\{z_i\}$. But σ_x maps $\{z_i\}$ into a single point of S. Hence Φ_x is single-valued. Distinct points in U_x have inequivalent images in S_x^* and therefore map into different points of D. That is, Φ_x is univalent and therefore one-to-one. Since τ_x is analytic (regular in all of S_x^*), it is open† and continuous there.

(3) $x \in \mathrm{P}$; define‡

$$\tau_x(z) = e\left(\frac{1}{c(z - x)}\right)$$

where, as usual,

$$z' = Pz, \qquad \frac{1}{z' - x} = \frac{1}{z - x} + c.$$

Here τ_x maps S_x^* onto an open disk D about the origin, the point x going into the origin. The mapping is ∞ to 1; the inverse image of $\tau_x(z) = w \neq 0$ is $\{P^m z,\ m = \text{integer}\}$, whereas 0 corresponds only to x. Again σ_x maps $\{P^m z\}$ on a single point of S. It follows as before that Φ_x is one-to-one.

Since τ_x is regular in $S_x^* - x$, it is open and continuous there. It is easy to see directly that τ_x is still continuous at $z = x$. To prove it is open on all

† We are using the well-known fact that a function regular in a region provides an open mapping of the region. This can be proved, for example, by means of Rouché's theorem.

‡ We are assuming x finite; otherwise we set

$$\tau_x(z) = e\left(\frac{z}{\lambda}\right).$$

of S_x^* it is sufficient to show that τ_x carries a basis set containing x (that is, an S_x) into an open set. Indeed, this is the case and we conclude that τ_x is open and continuous on S_x^*.

We have shown in each case that Φ_x is a homeomorphism of U_x onto an open disk of the complex plane. Thus S is a surface.

1G. We must finally show that the neighbor relations $\Phi_x \circ \Phi_y^{-1}$ are analytic functions.

Suppose Φ_x, Φ_y are homeomorphisms of neighborhoods $U_x = \sigma(S_x^*)$, $U_y = \sigma(S_y^*)$, respectively, and let $U_x \cap U_y = D' \neq \emptyset$. Select $x, y \in \mathrm{H}^+$ so that S_x^* intersects S_y^* in $D \neq \emptyset$. The equation

$$\Phi_x \circ \Phi_y^{-1} = \tau_x \circ \sigma_x^{-1} \circ \sigma_y \circ \tau_y^{-1} = \tau_x \circ \tau_y^{-1},$$

with domain $\Phi_y(D') = \tau_y(D)$, can be verified by a consideration of cases depending on the character of x and y. Note, for example, that the definition of S_x^* and S_y^* makes it impossible that D should contain distinct Γ-equivalent points unless they are mapped into the same point by τ_x.

From the definition of $\tau_x(z)$ in the various cases we observe that $\tau_x(z)$ is conformal except possibly at $z = x$, and $\tau_x^{-1}(w)$ is conformal except possibly at $w = 0$. But we have chosen S_x^*, S_y^* in such a way that D contains no fixed points. Hence $\tau_y(D)$ never contains the origin and so τ_y^{-1} is conformal on this set. The domain of τ_x is D and τ_x is conformal there. Hence $\Phi_x \circ \Phi_y^{-1}$ is analytic on its domain as required.

We have proved the following result:

THEOREM. Let Γ be a real discrete group. Then $S = \Gamma\backslash\mathrm{H}^+$, together with the structure (U_x, Φ_x) induced on it by the projection map σ, is a Riemann surface.

1H. An important class of Riemann surfaces are the *compact* surfaces. It is known that every compact Riemann surface is homeomorphic to a sphere with g handles, where $g \geq 0$ is called the genus. For $g = 0$ we have a sphere; for $g = 1$, a torus, etcetera. The next theorem gives a criterion for compactness in terms of the normal polygon of Γ.

THEOREM. The Riemann surface $S = \Gamma\backslash\mathrm{H}^+$ is compact if and only if the normal polygon $\bar{N}_0$, considered as a subset of H^+, is compact.

Since

$$\sigma(\bar{N}_0) = S,$$

the compactness of $\bar{N}_0$ and the continuity of σ imply that S is compact.

Before discussing the converse, let us note that the triangular region Δ cut out of N_0 at a parabolic vertex p by a horocycle through p is compact in the topology of H^+. Indeed, any covering of $\bar{\Delta}$ must contain a set S_p; the remainder $\bar{\Delta} - S_p$ is obviously compact. From any covering of $\bar{\Delta}$, therefore, we select S_p and a finite subcovering of $\bar{\Delta} - S_p$, which together form a finite covering of $\bar{\Delta}$.

The existence of a free side in N_0 automatically excludes the possibility that $\bar{N}_0$ is compact. For $\bar{N}_0$ does not include the inner points of a free side, since such points do not belong to H^+. But if N_0 has no free sides and has a finite number of sides (in H), then $\bar{N}_0$ is compact. For then N_0 has a finite number of parabolic cusps (I, Theorem 41) and the preceding argument applies. N_0 is composed of a relatively compact set in H and a finite number of triangular regions Δ described above, whose closures are compact in H^+.

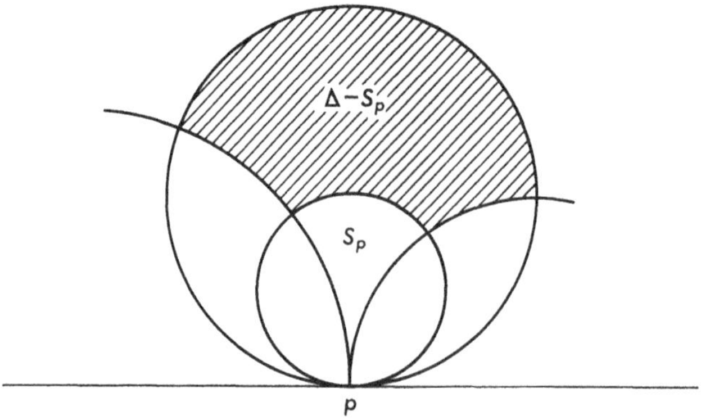

FIG. 28.

We return to the proof of the theorem and suppose that S is compact. We shall show that N_0 has a finite number of sides and no free sides. Suppose first that N_0 has an infinite number of sides. Then we can choose an infinite set of inequivalent points $\{z_i\}$, each being an inner point of a different side of N_0. The points $q_n = \sigma(z_n) \in S$ are therefore all distinct, and because of the compactness of S, possess an accumulation point $q \in S$. Let $\zeta \in H^+$ be an inverse image of q; that is, $\sigma(\zeta) = q$. Since every neighborhood D of ζ maps into a neighborhood of q, we can assert that D contains a Γ-image of infinitely many z_i. If we choose D small enough, it will follow that ζ is not a point of $\bar{N}_0 \cap H$, for D will contain the equivalent of at most one z_i, since z_i is not a vertex. Thus ζ is a parabolic vertex p. A small enough D intersects only the two sides of N_0 meeting at the vertex† and D can be made still smaller, if necessary, to exclude the images of the finite number of z_i lying on the sides of N_0 issuing from the points of the parabolic cycle determined by p. We

† In this case D is the interior of a horocycle at p (plus the point p) and meets infinitely many sides of the network of normal polygons. But only two of those sides belong to N_0.

conclude that $\zeta \notin \mathrm{H}^+$, which is a contradiction. Hence N_0 has a finite number of sides.

Second, if N_0 has a free side f, let $z_i \to x \in \operatorname{Int} f$ with $z_i \in N_0$. Then $q_i = \sigma(z_i) \to q$, a point of S, and $\zeta \in \mathrm{H}^+$ where $\sigma(\zeta) = q$. By the reasoning used above we obtain a contradiction and can conclude that N_0 has no free sides.

We have now shown that N_0 has a finite number of sides and no free sides and it follows that $\bar{N}_0$ is compact in H^+.

As an example consider the modular group. Its fundamental region N_0 is not compact in H but is compact in H^+. Our construction may be regarded as a compactification of the Riemann surface $\Gamma\backslash\mathrm{H}$, which has a hole at the image point of the parabolic vertex $\{i\infty\}$ in N_0 (see Exercises 1 and 2).

Our results show that we can certainly make this compactification if N_0 has a finite number of sides (and therefore a finite number of parabolic vertices) and no free sides. What if N_0 has infinitely many sides? Then the preceding argument shows $\bar{N}_0$ is not compact in H^+. Indeed, any accumulation point z^* of the set $\{z_i\}$ must lie on E but cannot be a parabolic vertex, for its every neighborhood meets infinitely many sides of N_0. Thus $z^* \notin \mathrm{H}^+$ and $\bar{N}_0$ is not compact. We have already seen that this is also true if N_0 has a free side. *The surface $\Gamma\backslash\mathrm{H}^+$ is a Riemann surface, but a noncompact one, when N_0 has infinitely many sides or has a free side.*

Note: If N_0 has a finite number of sides and a finite (positive) number of free sides, the Riemann surface $\Gamma\backslash\mathrm{H}^+$ is often called a *compact surface with boundary*. It can be obtained from a true compact surface by deleting a finite number of points and disks.

Exercise 1. Let $S' = \Gamma\backslash\mathrm{H}$ be the orbit space of H, rather than H^+, under the mappings of Γ. We proceed as before, but all mappings, etcetera, are restricted to H. For homeomorphisms of neighborhoods in S' use those Φ_x of the text whose domains lie in S'. Show S' is a Riemann surface.

Exercise 2. S' is compact if and only if N_0 is relatively compact in H. [Cover $\bar{N}_0 \cap \mathrm{H}$ by open disks lying in H, map into S', select a finite subcovering, and map back in H.]

1I. The only topological invariant of a compact Riemann surface $S = \Gamma\backslash\mathrm{H}^+$ is its *genus* g. We shall now reveal how g can be computed from the normal polygon of Γ.

Since we are assuming S compact, the normal polygon N_0 will have a finite number of sides. We triangulate N_0 by selecting any interior point P and drawing the H-lines connecting P to each vertex. Since we are operating on the complex sphere, there is no exception to this procedure if one of the vertices should happen to be ∞.

The polygon N_0 is now a complex with a finite number of vertices, sides, and triangles. Certain sides are identified by the mappings of Γ, and also certain vertices. We can now apply Euler's formula:

$$\alpha_0 - \alpha_1 + \alpha_2 = 2 - 2g,$$

where α_0, α_1, α_2 are the number of vertices, sides, and triangles, respectively, identified elements being counted as one.

The integer α_0 is clearly one more than the number of different cycles† in N_0, while α_1 is $3n$, with n half the number of sides of N_0. Also $\alpha_2 = 2n$. Thus if c is the number of cycles,

$$1 + c - 3n + 2n = 2 - 2g,$$

or

$$c - n + 1 = 2 - 2g. \tag{1}$$

For the modular group we get

$$3 - 2 + 1 = 2 - 2g,$$

or $g = 0$. This can also be seen by looking at the normal polygon and making the identifications.

We shall now prove that the number g obtained from (1) is the genus of $S = \Gamma\backslash\mathrm{H}^+$, and for this purpose it is sufficient to establish a homeomorphism between S and N_0^*, where N_0^* is the closed polygon N_0 with equivalent sides and vertices identified. We may also regard N_0^* as the set obtained by adjoining certain boundary points to N_0—namely, exactly one vertex from each cycle and exactly one side from each pair of conjugate sides. In fact N_0^* is the polygon we used previously to compute the genus.

We wish to assign local structures to N_0^* and S under which they will be, first, manifolds, and second, homeomorphic. The structure already assigned to S will not be changed. In N_0^* we define

$$T_x = \Gamma S_x \cap N_0^*, \qquad x \in N_0^*.$$

It is seen that $\{T_x\}$ forms a basis for a topology for N_0^*. The resulting neighborhoods are shown in the figures; equivalent sides and vertices must of course be identified. Note particularly that N_0^* is locally compact; the parabolic vertices cause no difficulty. We can go on to define homeomorphisms of these neighborhoods in the same way we did for $S = \Gamma\backslash\mathrm{H}^+$. This structure makes N_0^* a manifold.

Now let

$$\sigma_1 = \text{restriction of } \sigma \text{ to } N_0^*.$$

† P is included among the vertices.

Since N_0^* is a fundamental set, σ_1 is one-to-one onto S. To complete the proof we would have to show that σ_1 is open and continuous. We do not give the details because they are almost the same as in the previous case.

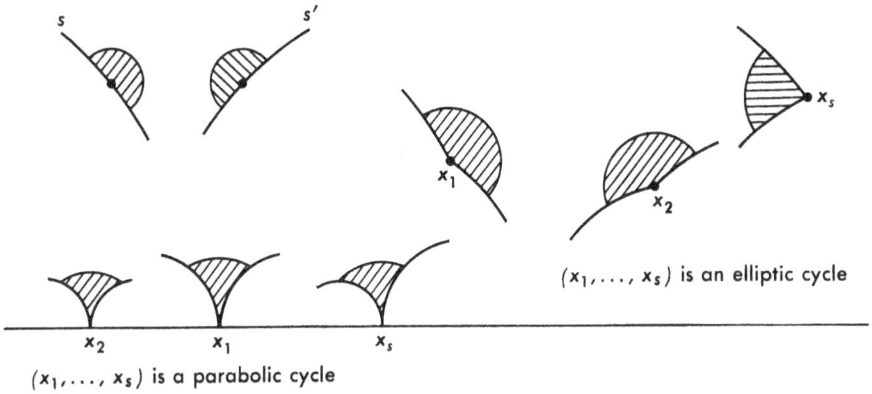

FIG. 29.

THEOREM. The genus g of the compact Riemann surface $\Gamma \backslash \mathrm{H}^+$ is given by

$$c - n + 1 = 2 - 2g,$$

where c is the number of cycles and $2n$ the number of sides of a normal polygon of Γ.

It follows incidentally that $c - n$ is an invariant for all normal polygons, even though c and n individually are not.

Exercise 1. Show that an elliptic or parabolic cyclic group has genus 0, and a hyperbolic cyclic group has genus 1.
[These surfaces are compact with boundary, but the genus can still be computed by the above theorem.]

2. Functions and Differentials

In this section we shall define functions and differentials on the Riemann surface $S = \Gamma \backslash \mathrm{H}^+$ and discuss their relationships to automorphic functions and forms on Γ in the domain H^+.

2A. Let us recall that any open subset U of S that contains q and is the domain of a homeomorphism Φ is called a parametric neighborhood of q. If in particular $\Phi(U)$ is a euclidean disk, U is called a *parametric disk* and $\Phi^{-1}(t^*)$ is called the *center* of U, where t^* is the center of $\Phi(U)$. We also call an open set V a parametric disk if $V \subset U$, where U is a parametric neighborhood and $\Phi(V)$ is a disk.

LEMMA. Each $q \in S$ is the center of a parametric disk.

Let $q \in U$ and let $\Phi(q) = t^*$. Define $D: |t - t^*| < r$, where r is so small that D lies in $\Phi(U)$. Then $K = \Phi^{-1}(D)$ is a parametric disk with center q.

Let t, τ be local variables of a point q; that is, $t = \Phi(p)$, $p \in U$ and $\tau = \Psi(p)$, $p \in V$, where $q \in U$, $q \in V$. Then $t = \Phi \circ \Psi^{-1} \circ \Psi(p) = \Phi \circ \Psi^{-1}(\tau)$, so t is an analytic function of τ. Hence

$$t - t_0 = \alpha_1(\tau - \tau_0) + \alpha_2(\tau - \tau_0)^2 + \cdots, \qquad \alpha_1 \neq 0, \tag{2}$$

where $t_0 = t(q)$, $\tau_0 = \tau(q)$.

A *function* φ on S is a mapping from S to the complex numbers. In each parametric disk U, φ becomes a function in the usual sense of the local variable t:

$$\varphi(q) = \varphi(\Phi^{-1}(t)) = \hat{\varphi}(t), \qquad t = \Phi(q), \qquad q \in U.$$

We define φ to be *meromorphic* on S if φ is identically zero or if in each parametric disk U there is an expansion

$$\varphi(q) = \hat{\varphi}(t) = a_\mu t^\mu + a_{\mu+1}t^{\mu+1} + \cdots, \qquad a_\mu \neq 0, \qquad \mu < \infty$$

for $q \in U$, where t is a variable on U that maps the center of U into $t = 0$. Had we used a different local variable τ, we would have

$$\begin{aligned}\hat{\varphi}(t) &= a_\mu(\alpha_1\tau + \alpha_2\tau^2 + \cdots)^\mu + a_{\mu+1}(\alpha_1\tau + \cdots)^{\mu+1} + \cdots \\ &= b_\mu\tau^\mu + \cdots = \varphi^*(\tau), \qquad b_\mu \neq 0.\end{aligned}$$

Thus μ is invariant with respect to changes in the local variable. If $\mu = 0$ and q_0 is the center of U,

$$\varphi(q_0) = \hat{\varphi}(0) = a_0 = b_0 = \varphi^*(0).$$

We call $\mu = \eta(q_0, \varphi)$ the order of φ at q_0; φ has a zero of order μ if $\mu > 0$ or a pole of order $-\mu$ if $\mu < 0$. To the function $\varphi(q) \equiv 0$ we assign the order $+\infty$. Then, as usual,

$$\begin{aligned}\eta(q, \varphi_1\varphi_2) &= \eta(q, \varphi_1) + \eta(q, \varphi_2), \\ \eta(q, \varphi_1 + \varphi_2) &\geq \min\{\eta(q, \varphi_1), \eta(q, \varphi_2)\}.\end{aligned} \tag{3}$$

The number a_0 is called the value of φ at q_0. The order and value of a function are independent of the local variable used.

For short we shall write *function on S* instead of meromorphic function on S. The set of all functions on S with the customary addition and multiplication is a field over the complex numbers.

2B. Next we define differentials on S. A (*meromorphic*) *differential of weight m* (m = integer) is a rule that assigns to each point $q \in S$ and to each local variable t at q a meromorphic function $\psi(t)$. The assignment shall be such that if t_1 is another local variable at q and ψ_1 is the corresponding function, we have

$$\psi_1(t_1)(dt_1)^m = \psi(t)(dt)^m.$$

We denote a differential by $d\Omega$ or $d\Omega_m$ and write

$$d\Omega = \psi(t)(dt)^m,$$

meaning that locally $d\Omega$ is given by this expression. The order of $d\Omega$, written $\eta(q, d\Omega)$, is defined to be the order of ψ and so, by 2A, is a conformal invariant. The differentials of weight 0 are the functions on S while those of weight 1 are the usual differentials.

If φ is a function on S, we quickly verify that $d\Omega = \varphi'(t)\, dt$ is a differential of weight 1, where $\varphi = f \circ \Phi^{-1}(t)$, $t = \Phi(q)$. Indeed, if t_1 is another local variable, $t_1 = \Phi_1(q)$, and $\varphi_1 = f \circ \Phi^{-1}(t)$ is the corresponding function, then

$$\varphi_1'(t_1)\, dt_1 = \left(\frac{d\varphi}{dt}\frac{dt}{dt_1}\right) dt_1 = \varphi'(t)\, dt.$$

The mth power of a differential of weight 1 is a differential of weight m. If $d\Omega$ is a differential, so is $\varphi\, d\Omega$, φ being a function. The differentials of a given weight form a complex vector space. The product of two differentials of weights m_1 and m_2 is a differential of weight $m_1 + m_2$, and the quotient is of weight $m_1 - m_2$ provided the denominator is not identically zero. In particular, the quotient of two differentials of the same weight is a function.

2C. *From now on we assume S is compact* unless otherwise stated.

THEOREM 1. If $\varphi \not\equiv 0$,

$$\sum_{q \in S} \eta(q, \varphi) = 0.$$

This theorem will be assumed from Riemann surface theory. It says that a function on S has the same number of zeros as it has poles if we count them in correct multiplicity. *The number of poles is finite*, otherwise the compactness of S would produce a point of accumulation of poles and φ could not be meromorphic at that point. As corollaries of this theorem we get:

THEOREM 2. A nonconstant function on S has at least one pole.

Let $N(\varphi)$ be the number of poles of φ; that is,

$$N(\varphi) = \sum_{\eta<0} \eta(q, \varphi).$$

THEOREM 3. A nonconstant function φ on S assumes each complex value $N(\varphi)$ times.

Theorems 2 and 3 follow from Theorem 1 in exactly the same way as the corresponding theorems on automorphic functions (see II, 2E, Exercise 1; 2F).

2D. It is a fundamental result of Riemann surface theory that nonconstant functions and nonzero differentials exist on compact Riemann surfaces (Springer, Chapter 8). There is a famous theorem, valid for compact surfaces, that says something about the number of essentially different functions and differentials. In order to explain this theorem we must introduce *divisors.*

Divisors on a Riemann surface are defined in the same way as on the domain H^+ (see II, 2E). A divisor is a formal finite sum

$$D = \sum_{q\in S} u(q) \cdot q,$$

where $u(q)$ is an *integer* (rather than merely a rational number, as previously), and $u(q) = 0$ except for finitely many q. The divisor of a function f on S is

$$D(f) = \sum \eta(q, f) \cdot q;$$

as we remarked in 2C, the integer $\eta(q, f)$ is different from zero for only a finite number of q because of the compactness of S. Principal divisors, divisor classes, degree of a divisor, and degree of a divisor class have the same definitions as previously. From 2C, Theorem 1 we see that

$$\deg D(f) = 0$$

for every function f on S—that is, for every principal divisor.[2] Obviously

$$\deg D(\varphi_1\varphi_2) = \deg D(\varphi_1) + \deg D(\varphi_2), \qquad \deg D\left(\frac{1}{\varphi}\right) = -\deg D(\varphi).$$

We call D_1 a multiple of D_2 provided the order of D_1 is at least equal to the order of D_2 at every point of S. We say that the function $\varphi \not\equiv 0$ *is a multiple of* D if $D(\varphi)$ is a multiple of D. It follows that $\deg D(\varphi) \geq \deg D$ if φ is a multiple of D.

It is clear because of (3) that the set of multiples of a given divisor D forms a complex vector space, which is denoted by $\{D\}$. The dimension of $\{D\}$, written $\dim D$, is finite. Indeed, suppose D has negative order $-m$ at the point q_0. Every function in $\{D\}$ then has a principal part at q_0 consisting of at most m terms. Let us select k linearly independent functions in $\{D\}$ and attempt to construct a linear combination of them that will be zero at q_0. This leads to a system of $m + 1$ linear equations in k unknowns. If $\dim D = \infty$, we can choose $k > m + 1$ and the linear system will have a nontrivial solution. Since the number of poles in D is finite, it is now clear that we can construct a nontrivial linear combination φ of $\{\varphi_1, \cdots, \varphi_K\}$, $\varphi_i \in \{D\}$, which has no poles at all and so is a constant (2C, Theorem 2); in fact $\varphi \equiv 0$ since $\varphi(q_0) = 0$. This says that the functions $\{\varphi_1, \cdots, \varphi_K\}$ are linearly dependent, but we chose them as linearly independent functions.

Moreover, dim D depends only on the divisor class of D. Suppose D_1, D_2 are in this class; then $D_1 - D_2 = D_0$, a principal divisor. Let φ be a function whose divisor is D_0, and let φ_1, φ_2 be multiples of D_1, D_2 respectively. Then $\varphi_2\varphi$ is a multiple of $D_2 + D_0 = D_1$ and φ_1/φ is a multiple of $D_1 - D_0 = D_2$. The mapping χ: $\varphi_2 \to \varphi_2\varphi$ is from $\{D_2\}$ into $\{D_1\}$ and has an inverse χ^{-1}: $\varphi_1 \to \varphi_1/\varphi$ from $\{D_1\}$ into $\{D_2\}$. Hence χ is one-to-one from $\{D_2\}$ onto $\{D_1\}$ and is clearly an isomorphism. The two vector spaces $\{D_1\}$ and $\{D_2\}$ have the same dimension, which may thus be regarded as the dimension of the divisor class in which D lies.

The following conclusions are now evident. Since a multiple of the divisor 0 is everywhere regular, and since an everywhere regular function is a constant, $\dim 0 = 1$. Moreover $\deg D > 0$ implies $\dim D = 0$. In fact, for any $\varphi \in \{D\}$ we have $D(\varphi) \geq \deg D > 0$, but $\deg D(\varphi) = 0$ because φ is a function.

2E. Riemann proved the following inequality:

RIEMANN'S INEQUALITY. *If g is the genus of a Riemann surface S, then*

$$\dim D \geq -\deg D + 1 - g$$

for every divisor D.

From this theorem, for the proof of which we refer the reader to the last chapter of Springer, we can deduce several noteworthy results.

THEOREM 1. *On a compact Riemann surface S of genus g there exist nonconstant functions of valence $\leq g + 1$.*

Let $D = -l \cdot q_0$, where $q_0 \in S$ and $l > 0$. Then

$$\dim D \geq l + 1 - g.$$

If we choose $l = g + 1$, we have $\dim D \geq 2$; that is, there are at least two linearly independent functions having a pole of order at most $g + 1$ at q_0 and regular elsewhere. One of these may certainly be taken as a constant, but then the other one must be nonconstant.

COROLLARY. *There is a function that has a pole at any preassigned point of S and is otherwise regular.*

The nonconstant function of Theorem 1 is one such.

THEOREM 2. There exists a univalent function on S if and only if S is of genus zero.

Let $g = 0$. By Theorem 1 there are nonconstant functions of valence ≤ 1—that is, univalent functions.

Let S support a univalent function φ. In the neighborhood of each point of S, φ is analytic and one-to-one; hence φ and φ^{-1} are continuous there. That is, φ is a homeomorphism of S onto the complex sphere, so S is of genus zero.

2F. We can extend the notion of divisor to differentials. If $d\Omega \not\equiv 0$ is a differential, we define $D(d\Omega)$ to be the divisor whose order at each point is the order of $d\Omega$ at that point. The divisors of all differentials of weight m are in the same divisor class, since the quotient of two such differentials is a function. In particular, when $m = 1$ this divisor class is called the canonical class and is denoted by Z. It is known that

$$\deg Z = 2g - 2.$$

It follows that if Z_m denotes the class of differentials of weight m,

$$\deg Z_m = m(2g - 2).$$

For certainly $\deg (d\Omega_1)^m$ has this value and $D(d\Omega_1)^m$ is in Z_m.

Let us write $d\Omega$ for a differential of weight 1. We shall say that $d\Omega$ is a multiple of D if $D(d\Omega)$ is a multiple of D. By the reasoning used above for functions we can conclude that $d\Omega$ is a multiple of D if and only if the function $d\Omega/d\Omega_0$ is a multiple of $D - Z_0$, where $d\Omega_0$ is a fixed differential of weight 1 with divisor Z_0. Moreover, $\dim (D - Z_0) = \dim (D - Z)$ is uniquely defined, for the dimension depends only on the divisor class (see 2D).

Thus $\dim (-Z) = \dim (0 - Z)$ is the number of linearly independent differentials that are multiples of 0—that is, that are everywhere regular. Such differentials are called *differentials of the first kind* and it is proved in Riemann Surface theory that there are g of them. That is,

$$\dim (-Z) = g.$$

We can now state the fundamental

RIEMANN-ROCH THEOREM.

$$\dim D = -\deg D + 1 - g + \dim (-D - Z).$$

This is a relation between the number of linearly independent *functions* that are multiples of D and the number of linearly independent *differentials* that are multiples of $-D$.

In the Riemann-Roch theorem take $D = -Z$:

$$\dim(-Z) = \deg Z + 1 - g + \dim 0 = \deg Z + 2 - g.$$

This relation is consistent with the values of $\dim(-Z)$ and $\deg Z$ already given.

2G. THEOREM.† Let $K = K(S)$ be the field of meromorphic functions on S and $\mathbf{K} = \mathbf{K}(\Gamma)$ the field of automorphic functions on Γ, where $S = \Gamma\backslash H^+$. Then K is isomorphic to $\mathbf{K}$ by the mapping $f \leftrightarrow \varphi, f \in \mathbf{K}, \varphi \in K$, where

$$f(x) = \varphi(q), \qquad x \in H^+, \qquad q = \sigma x \in S \tag{4}$$

and σ is the projection mapping $H^+ \to S$. Also

$$n(x, f) = \eta(q, \varphi) \tag{5}$$

and

$$N(f) = N(\varphi).$$

Let $\varphi \in K$; then $f(x) = \varphi(\sigma x)$ is defined for $x \in H^+$ and

$$f(Vx) = \varphi(\sigma(Vx)) = \varphi(\sigma x) = f(x), \qquad V \in \Gamma$$

showing the invariance of f on Γ. Let ζ be a local variable at $q \in S$ with $\zeta(q) = 0$; for p near q we have

$$\varphi(p) = \hat{\varphi}(\zeta) = \sum_{n=\mu}^{\infty} a_n \zeta^n$$

with μ independent of the choice of ζ. We choose for ζ the variable $\Phi_x = \tau_x \circ \sigma_x^{-1}$ defined in 1F. Then $\zeta = \Phi_x(p) = \tau_x(z)$, z near x, is the proper local variable for the expansion of an automorphic function (II, 2C). Hence f is meromorphic on H^+ and therefore $f \in \mathbf{K}$. Moreover (5) holds.

Conversely, if $f \in \mathbf{K}$, then $\varphi(q) = f(\sigma^{-1}(q))$ is single-valued because f takes the same value at points of the orbit $\sigma^{-1}(q)$. The argument about the expansions is reversible so $\varphi \in K$.

The mapping $f \leftrightarrow \varphi$ is obviously an isomorphism, hence $K \cong \mathbf{K}$.

By definition there are $N(f)$ poles of f in a fundamental set R^*. Since σ is clearly one-to-one from R^* onto S and we have just seen that polar orders correspond, the function φ must have $N(f)$ poles. This concludes the proof.

COROLLARY. There exists a univalent function on Γ if and only if the genus of Γ is zero.

This is an immediate consequence of 2E, Theorem 2.

† This theorem does not require S to be compact.

2H. We extend the above results to automorphic forms.

THEOREM[†]. Let H_m be the vector space of meromorphic differentials of weight m on S; $\{\Gamma, -2m\}$ is, as usual, the vector space of automorphic forms on Γ of dimension $-2m$. Then $\{\Gamma, -2m\}$ is isomorphic to H_m by $F \leftrightarrow d\Omega$, where

$$F(x)(dx)^m = d\Omega(q), \qquad q = \sigma x, \qquad x \in \mathrm{H}^+ \tag{6}$$

and

$$n(x, F) = \eta(q, d\Omega) + m\left(1 - \frac{1}{l}\right) \tag{7}$$

where $l = l(x)$ is 1 if x is a nonfixed point, l is the order of x if x is an elliptic fixed point, and $l = \infty (1/l = 0)$ if x is a parabolic fixed point.

To prove the theorem observe first that (6) shows $F(x)(dx)^m$ to be invariant under Γ if $d\Omega$ is given. Let $d\Omega = g(\zeta)(d\zeta)^m$ locally; g is meromorphic and we have, for small ζ,

$$g(\zeta) = \sum_{n=\mu}^{\infty} a_n \zeta^n.$$

Now $d\Omega$ is by definition invariant to changes of local variable, so we use for ζ the variable $\Phi_x = \tau_x(z) = t$, z near x, of 1F. This gives, after some calculation,

$$F(z) = g(t)\left(\frac{dt}{dz}\right)^m = \begin{cases} a_\mu t^\mu + \cdots, & x = \text{nonfixed point} \\ (z - \bar{x})^{-2m} t^{-m/l}\{b_\mu t^{m+\mu} + \cdots\}, & \\ \quad x = \text{elliptic point of order } l & \\ (z - x)^{-2m} t^m \{c_\mu t^\mu + \cdots\}, & \\ \quad x = \text{parabolic vertex.} & \end{cases}$$

Since t is the correct local variable for an automorphic form (see II, 2C), we can assert that F is meromorphic in H^+, and (7) follows at once.

Conversely, suppose $F \in \{\Gamma, -2m\}$ is given. Then $d\Omega(q) = F(\sigma^{-1}q)(d\sigma^{-1}q)^m$ is single-valued on S because of the invariance of the right member on the orbit $\sigma^{-1}q$. We get the expansions for $d\Omega$ from those of F by reversing the above argument. Hence $d\Omega \in H_m$. That the mapping $F \leftrightarrow d\Omega$ is an isomorphism is obvious, and the proof is complete.

2I. Between the divisors of $F \in \{\Gamma, -2m\}$ and $d\Omega \in H_m$ satisfying (6) there is the relation

$$D(F) = D(d\Omega) + \tilde{D}, \tag{8}$$

where $\tilde{D}$ is the divisor on H^+ defined by

$$\tilde{D} = \sum_{x \in N^*} m\left(1 - \frac{1}{l}\right) \cdot x, \qquad l = l(x).$$

[†] This theorem does not require S to be compact.

This is merely a rewriting of (7). Suppose $d\Omega$ is a differential of weight 1; then $F \in \{\Gamma, -2\}$ and

$$\deg D(d\Omega) = \deg D(F) - \deg \tilde{D}.$$

We take $\deg D(F)$ from II, Theorem 3D and get

$$\deg D(d\Omega) = 2g - 2 + \sum\left(1 - \frac{1}{l}\right) - \sum\left(1 - \frac{1}{l}\right) = 2g - 2,$$

a result stated without proof in 2F.

Notice that a differential of the first kind on S (that is, an everywhere regular differential) corresponds to a *cusp form of dimension* -2.

2J. We are going to use the Riemann-Roch theorem to calculate the dimension of $C^+ = C^+(\Gamma, -2m)$, the subspace of $\{\Gamma, -2m\}$ consisting of everywhere regular automorphic forms.

First, let $m = 0$. Then $C^+(\Gamma, 0)$ is isomorphic to the space of everywhere regular functions on S. All such functions being constants, we have

$$\dim C^+(\Gamma, 0) = 1.$$

Now suppose $m \neq 0$. If $F_0 \not\equiv 0$ is a fixed form in $\{\Gamma, -2m\}$ and $F \in C^+(\Gamma, -2m)$, then $F/F_0 = f$ is an automorphic function such that fF_0 is regular everywhere. Conversely, if f is an automorphic function having this property, then $fF_0 = F \in C^+$. The mapping $F \to f$ is an isomorphism of $C^+(\Gamma, -2m)$ and the space

$$\mathbf{L} = \{f \in \{\Gamma, 0\} \mid fF_0 \text{ is regular in } \mathrm{H}^+\}.$$

Hence we have to calculate $\dim \mathbf{L}$.

The condition $f \in \mathbf{L}$ is tantamount to

$$n(z, f) + n(z, F_0) \geq 0, \qquad z \in \mathrm{H}^+.$$

On S this becomes, with $f \leftrightarrow \varphi$, $F_0 \leftrightarrow d\Omega_0$, $q = \sigma x$,

$$\eta(q, \varphi) + \eta(q, d\Omega_0) + m\left(1 - \frac{1}{l}\right) \geq 0,$$

with the usual definition of $l = l(q)$. Set up the divisor

$$D^* = \sum_{q \in S}\left[m\left(1 - \frac{1}{l}\right)\right] \cdot q,$$

with $[u]$ the greatest integer $\leq u$. Since $u \geq 0$ implies $[u] \geq 0$ and conversely, we see that $f \in \mathbf{L}$ if and only if φ is a multiple of

$$D_1 = D(-d\Omega_0) - D^*.$$

Hence

$$\dim C^+ = \dim \mathbf{L} = \dim D_1.$$

Since $d\Omega_0$ has weight m, we get

$$\deg D_1 = -2m(g-1) - \sum \left[m\left(1 - \frac{1}{l}\right)\right].$$

The sum in the right member, as well as all subsequent ones, may be regarded as extending either over S or over a fundamental set N^*.

(1) Let $m < 0$. Since $-[x] \geq -x$,

$$\deg D_1 \geq -2m \left\{ g - 1 + \frac{1}{2} \sum \left(1 - \frac{1}{l}\right)\right\} > 0,$$

for the expression in braces is proportional to the hyperbolic area of the normal polygon. By the remark at the end of 2D, $\dim D_1 = 0$ and it follows that

$$\dim C^+(\Gamma, -2m) = 0, \qquad m < 0.$$

That is, there are no nonzero everwhere regular automorphic forms of even positive dimension. This is a result we obtained in Chapter II, Theorem 3E.

(2) Let $m > 0$. By the Riemann-Roch theorem

$$\dim D_1 = -\deg D_1 + 1 - g + \dim(-D_1 - Z).$$

Now

$$\deg(-D_1 - Z) = 2(m-1)(g-1) + \sum \left[m\left(1 - \frac{1}{l}\right)\right]. \tag{9}$$

LEMMA. For $m \geq 1$, $l \geq 1$ we have

$$\left[m\left(1 - \frac{1}{l}\right)\right] \geq (m-1)\left(1 - \frac{1}{l}\right).$$

With the obvious case $l = 1$ set aside, the result is equivalent to

$$\frac{m}{l} + \left[-\frac{m}{l}\right] \geq \frac{1}{l} - 1.$$

Since the left member is periodic in m with period l, we may assume $1 \leq m < l$. Then $[-m/l] = -1$ and the inequality reduces to

$$\frac{m}{l} \geq \frac{1}{l}.$$

Using the lemma in (9) we get

$$\deg(-D_1 - Z) \geq 2(m-1)\left\{ g - 1 + \frac{1}{2} \sum \left(1 - \frac{1}{l}\right)\right\},$$

and this is certainly positive if $m > 1$. Hence

$$\dim C^+(\Gamma, -2m) = -\deg D_1 + 1 - g$$
$$= 2m(g-1) + \sum\left[m\left(1-\frac{1}{l}\right)\right] - (g-1),$$

$$\dim C^+(\Gamma, -2m) = (2m-1)(g-1) + \sum\left[m\left(1-\frac{1}{l}\right)\right], \qquad m \geq 2. \quad (10)$$

When $m = 1$ we obtain from (9)

$$\deg(-D_1 - Z) = \sum\left[1-\frac{1}{l}\right] = \sigma_0,$$

where σ_0 is the number of inequivalent parabolic vertices in the closure of a fundamental region. Thus when $\sigma_0 > 0$ we again have $\dim(-D_1 - Z) = 0$ and the formula (10) holds. We have proved the following

THEOREM. Let Γ be a horocyclic group with compact Riemann surface $\Gamma\backslash H^+$. Then

$$\dim C^+(\Gamma, -2m) = \begin{cases} 1, & m = 0 \\ 0, & m < 0 \\ (2m-1)(g-1) + \sum\left[m\left(1-\frac{1}{l}\right)\right], \\ \qquad m > 0, \text{ provided } \sigma_0 > 0 \text{ when } m = 1 \end{cases}$$

The case we had to exclude is $m = 1, \sigma_0 = 0$. The everywhere regular forms in $\{\Gamma, -2\}$ include the cusp forms, and a cusp form corresponds to an everywhere regular differential on S. Since there are g of those, we have

$$\dim C^+(\Gamma, -2) \geq g.$$

Exercise 1. Prove there are nonconstant automorphic forms of *positive* dimension on Γ that have singularities only at parabolic vertices.
[The vector space of the desired forms is isomorphic to the space of functions f such that fF_0 is regular in H, where F_0 is an arbitrary nonzero form in $\{\Gamma, -r\}$, $r < 0$. Here we may allow f to have arbitrary polar singularities at parabolic cusps; choose the polar orders so large that the dimension of the space of f's is positive.]

2K. We apply these results to the modular group M. Its genus is zero. There is one parabolic class, one elliptic class of order 2, and one of order 3. Hence

$$\dim C^+(M, -2m) = -2m + 1 + m + \left[\frac{m}{2}\right] + \left[\frac{2}{3}m\right].$$

Set $m = 6k + r, 0 \leq r < 6, k = 0, 1, 2, \cdots$; then

$$\dim C^{+}(M, -2m) = k + 1 + \left\{-r + \left[\frac{r}{2}\right] + \left[\frac{2}{3}r\right]\right\}.$$

But $k = [m/6]$ and the expression in braces vanishes for $2 \leq r < 6$ and $r = 0$, and is -1 for $r = 1$. Therefore

$$\dim C^{+}(M, -2m) = \begin{cases} \left[\dfrac{m}{6}\right], & m \equiv 1 \pmod 6 \\ \left[\dfrac{m}{6}\right] + 1, & m \equiv 1 \pmod 6 \end{cases}$$

In particular there are no everywhere regular modular forms of dimension -2.

A particularly interesting case is $m = 6$. Here there are two linearly independent forms, and so some linear combination of them must vanish at $i\infty$. If we call this new form $\Delta(\tau)$, then Δ is a cusp form. Bringing in the Poincaré series $G_{-4}(\tau, 0)$, $G_{-6}(\tau, 0)$ of II, 1A, (3), we can assert that for suitable constants α, β we have

$$\alpha G^{3}_{-4} + \beta G^{2}_{-6} = \Delta.$$

By actual numerical control of the coefficients, we find

$$\Delta(\tau) = c_1 t + c_2 t^2 + \cdots, \qquad t = e(\tau)$$

with $c_1 \neq 0$. That is, Δ has a simple zero at $i\infty$. Now the expression for $\sum n(z, \Delta)$ developed in II, 3D shows that Δ has zeros of total order 1 (since it has no poles); *hence $\Delta(\tau)$ is zerofree in* H. Hence the function

$$J(\tau) = \frac{G^{3}_{-4}}{\Delta} = a_{-1}t^{-1} + a_0 + a_1 t + \cdots,$$

which, up to a constant factor, is Klein's absolute modular invariant, is an automorphic (modular) function that is regular in H and has a simple pole in the normal polygon at the parabolic cusp $i\infty$. It is determined uniquely by the coefficients a_{-1} and a_0, for the difference of two such functions would have a zero (at $i\infty$) but no pole in the closed fundamental region. It turns out that if we normalize by requiring $a_{-1} = 1$, then all a_n with $n > 0$ are (rational) integers, and they have many interesting arithmetical properties.

Exercise 1. Calculate $\dim C^{+}(\Gamma, -2m)$ for $\Gamma = \Gamma(2)$—see I, 6C.

2L. Since $S = \Gamma \backslash H^{+}$ has been assumed compact, the field $K(S)$ of meromorphic functions on S is an algebraic function field of one variable (Springer, page 289). That is, there exist functions φ, ψ on S such that $K(S)$

is the field of rational functions of φ, ψ. Suppose under the isomorphism $K(S) \to \mathbf{K}(\Gamma)$ the function φ goes into f, ψ goes into g. Let the arbitrary function $w \in \mathbf{K}(\Gamma)$ correspond to $\omega \in K(S)$. Since constants remain fixed under the isomorphism, the fact that $\omega = R(\varphi, \psi)$ implies that $w = R(f, g)$. Hence

THEOREM. If $\Gamma\backslash H^+$ is compact, $\mathbf{K}(\Gamma)$ is isomorphic to an algebraic function field of one variable.

2M. If Γ_1, Γ_2 are discrete groups and S_1, S_2 the corresponding Riemann surfaces, then S_1 is conformally equivalent to S_2 if and only if $\Gamma_1 = A\Gamma_2A^{-1}$ for an $A \in \Omega_R$. Thus each conformal class of Riemann surfaces is represented by a conjugacy class of discrete groups acting on H. This fundamental theorem has many beautiful and important consequences, some of which are sketched in Note 3 to this chapter (see below).

Notes to Chapter 3.

1. Given a set S we say **U** is a *topology for* S if **U** is a collection of subsets of S (called *open sets*) such that S and the null set are open, and arbitrary unions and finite intersections of open sets are open. A subcollection $\mathbf{B} \subset \mathbf{U}$ is a *basis for the topology* **U** provided every open set in **U** is a union of members of **B**.

2. The converse, however, is false. It is not necessarily the case that a function f exists having a given distribution of zeros and poles subject merely to the restriction that $\deg D(f) = 0$. A necessary and sufficient condition is given by *Abel's theorem* (see Springer, page 277).

3. The theory of conformal equivalence of Riemann surfaces is a most beautiful and elegant part of Riemann surface theory, as well as an important one. Let us begin our discussion by defining conformal equivalence.

Let S_1, S_2 be Riemann surfaces with homeomorphisms $\{\Phi_1\}$, $\{\Phi_2\}$, and let f map S_1 into S_2. Then $\Phi_2 \circ f \circ \Phi_1^{-1}$, when defined, is a mapping from one complex plane to another. When all possible functions of this type are analytic, we say f is an analytic mapping of S_1 into S_2. If in addition f is one-to-one, we say f is a conformal mapping (conformal homeomorphism). Two surfaces that can be mapped onto each other by a conformal mapping are called *conformally equivalent.*

We have seen that a discrete group Γ acting on H defines a Riemann surface $\Gamma\backslash H$ (see 1H, Exercise 1). But we can reverse the process. Starting from a given Riemann surface S we can find a group Γ such that $\Gamma\backslash H = S$.

To carry out this program we introduce some new concepts. Suppose R is a surface (not necessarily a Riemann surface). The pair $(\tilde{R}, \pi)$ is called a smooth covering surface of R provided $\tilde{R}$ is a surface and π is a locally topological mapping of $\tilde{R}$ into R. The points in the inverse image $\pi^{-1}(q)$, $q \in R$, are said to *lie over* q, and π is called the projection mapping. The surface $\tilde{R}$ is said to be *unlimited* if, given a curve γ on R with initial point q and a $\tilde{q} \in \tilde{R}$ lying over q, there is a curve $\tilde{\gamma}$ on $\tilde{R}$ starting from $\tilde{q}$ and lying over γ; that is, $\pi\tilde{\gamma} = \gamma$. We define a *covering transformation h* of $\tilde{R}$ to be a homeomorphism of $\tilde{R}$ on itself that carries a point into another point with the same projection ($\pi \circ h = \pi$). The set of all covering transformations forms a group.

Of all smooth unlimited coverings of R there is one, and essentially only one, that is *simply connected.* This surface is called the universal covering surface of R and is denoted by $\hat{R}$. Let $\hat{H}$ be the group of covering transformations of $\hat{R}$.

When we specialize to the case of a Riemann surface S with universal covering surface $\hat{S}$, the projection map π turns out to be an *analytic* mapping of $\hat{S}$ into S and $h \in \hat{H}$ is a *conformal* automorphism of $\hat{S}$. Since $\hat{S}$ is simply connected, we can now apply Riemann's Mapping Theorem: *every simply connected Riemann surface is conformally equivalent to the extended plane, the finite plane, or the upper half-plane.* Let us consider only the last possibility (hyperbolic case), which is the common one.

There is a function f that maps $\hat{S}$ conformally on H. The group $f \circ \hat{H} \circ f^{-1} = \Gamma$ is therefore a group of conformal homeomorphisms of H, in other words, linear transformations. It is easily shown that $\hat{H}$ is discontinuous and only the identity element of $\hat{H}$ has a fixed point; the same must therefore be true of Γ. Thus Γ is a discontinuous group of fixed-point-free linear transformations of H. Two points equivalent under Γ are carried by f^{-1} into two points of $\hat{S}$ that have the same projection in S; that is, $\pi \circ f^{-1}$ identifies points of H equivalent under Γ. This leads to a proof that $\Gamma\backslash\mathrm{H} = S$. We can identify $\pi \circ f^{-1}$ with the map σ: H $\to S$ of 1C.

If we replace Γ by $L\Gamma L^{-1}$, where $L \in \Omega_R$, this is equivalent to replacing f by Lf, still a conformal mapping of $\hat{S}$ on H. Thus $L\Gamma L^{-1}\backslash\mathrm{H}$ is conformally equivalent to $\Gamma\backslash\mathrm{H}$. Suppose conversely that φ maps $S_1 = \Gamma_1\backslash\mathrm{H}$ conformally on $S_2 = \Gamma_2\backslash\mathrm{H}$. Let $\sigma_i (i = 1, 2)$ be the mapping that identifies Γ_i-equivalent points. For $z_0 \in \mathrm{H}$ the point $\varphi^{-1} \circ \sigma_2(z_0)$ is in S_1 and we select a neighborhood of this point that is topologically equivalent by σ_1 to a neighborhood in H. Then we can define a restricted inverse σ_1^{-1} and $L = \sigma_1^{-1} \circ \varphi^{-1} \circ \sigma_2$ is analytic at z_0 and can be continued analytically throughout H. By the Monodromy theorem L is single-valued in H. We can show that L is one-to-one and onto, and so is a linear transformation. It is now easily checked that $\Gamma_1 = L\Gamma_2 L^{-1}$. Hence

THEOREM. Two Riemann surfaces $S_1 = \Gamma_1\backslash\mathrm{H}$ and $S_2 = \Gamma_2\backslash\mathrm{H}$ are conformally equivalent if and only if $\Gamma_1 = L\Gamma_2 L^{-1}$ with L a conformal homeomorphism of H.

Thus a conformal class of Riemann surfaces corresponds to a conjugacy class of discrete groups. The group Γ may be regarded as a normal form of the Riemann surface $\Gamma\backslash\mathrm{H}$.

Let us now go on to consider conformal homeomorphisms of S *on itself*; such mappings form a group $C(S)$. Suppose $S = \Gamma\backslash \mathrm{H}$ and σ is the map $\mathrm{H} \to S$. If $\varphi \in C(S)$, then according to the above argument $L = \sigma^{-1} \circ \varphi \circ \sigma$ is an element of Ω_R such that $\Gamma = L\Gamma L^{-1}$. In other words L lies in N, the normalizer of Γ in Ω_R. Since evidently LV gives rise to the same φ whenever $V \in \Gamma$, each φ is associated uniquely with a coset $L\Gamma$ of N/Γ. Hence $C(S) \cong N/\Gamma$.

Consider now a compact surface S of genus $g > 1$. It can be shown that $\hat{S}$ is hyperbolic ($\hat{S}$ conformally equivalent to H). In fact, suppose $\hat{S}$ equivalent to the finite plane. Then Γ can contain, besides the identity, only elements having the unique fixed point ∞—that is, translations. Hence Γ is either the identity or the simply or doubly periodic group, the corresponding S being the plane, the punctured plane, and the torus. By similar arguments it is shown that $\hat{S}$ cannot be equivalent to the extended plane.

It follows that Γ can contain only the identity and hyperbolic elements. For the fundamental region R of Γ is compact in H, as we found in 1H, Exercise 2; hence Γ has no parabolic elements (I, 4J, Theorem 2). But also Γ has no element with a fixed point in H—that is, no elliptic elements.

Now Γ is not abelian, for all abelian discrete groups on H are cyclic (I, 2G) and a hyperbolic cyclic group has genus 1 (see 1I, Exercise 1). It can be shown from this fact that N is discrete. But Γ is horocyclic and therefore so is N. Let $R(N)$, $R(\Gamma)$ be normal polygons of N and Γ. We denote by $|R|$ the hyperbolic area of R.

According to Siegel's theorem of I, 5D we know that $|R(N)| \geq \pi/21$. On the other hand formula (26) of I, 5D tells us that $|R(\Gamma)| = 4\pi(g-1)$, since Γ has no elliptic or parabolic elements. Thus

$$\mu = \frac{|R(\Gamma)|}{|R(N)|} \leq 84(g-1).$$

But μ is simply the index of Γ in N, for $R(\Gamma)$ consists of μ copies of $R(N)$, all with the same hyperbolic area (I, 6D). Since $C(S) \cong N/\Gamma$, we have

THEOREM (A. Hurwitz). If S is a compact Riemann surface of genus $g > 1$, $C(S)$ is a finite group of order at most $84(g-1)$.

The upper bound is clearly attained whenever N is the triangle group (2, 3, 7)—see end of I, 5D.

References

Lars Ahlfors and Leo Sario, *Riemann Surfaces*, Princeton University Press, Princeton, 1960

Pierre Fatou, *Fonctions automorphes* (vol. 2 of *Théorie des fonctions algébriques* . . . P. E. Appell and Édouard Goursat), Gauthiers-Villars, Paris, 1930.

L. R. Ford, *Automorphic Functions*, McGraw-Hill, New York, 1929; 2d ed., Chelsea, New York, 1951.

Robert Fricke and Felix Klein,

1. *Vorlesungen über die Theorie der Modulfunktionen*, vol. 1, 1890; vol. 2, 1892; Teubner, Leipzig.

2. *Vorlesungen über die Theorie der automorphen Funktionen*, vol. 1, 1897; vol. 2, pt. 1, 1901; vol. 2, pt. 2, 1912; Teubner, Leipzig.

R. C. Gunning, *Lectures on Modular Forms*, Annals of Mathematics Studies No. 48, Princeton University Press, Princeton, 1962.

J. Lehner, *Discontinuous Groups and Automorphic Functions*, American Mathematical Society, Providence, 1964.

R. Nevanlinna, *Uniformisierung*, Springer, Berlin, 1953

A. Pfluger, *Theorie der Riemannschen Flächen*, Springer, Berlin, 1957.

L. Schlesinger, *Automorphe Funktionen*, de Gruyter, Berlin, 1924.

C. L. Siegel, *Ausgewählte Fragen der Funktionentheorie* II, Mathematisches Institut der Universität Göttingen, Gottingen, 1954.

G. Springer, *Introduction to Riemann Surfaces*, Addison-Wesley, Reading, Mass., 1957.

H. Weyl, *Die Idee der Riemanschen Fläche*, 1st ed., Teubner, Berlin, 1913; 3d ed., 1955.

Glossary

Symbol	*Description*	*Page*
$(a\ b \mid c\ d)$	the matrix $\begin{pmatrix} a & b \\ c & d \end{pmatrix}$	
	automorphic form	76
	automorphic function	77
	bilinear mapping	3
$C^+\{\Gamma,-r\}, C^0\{\Gamma,-r\}$	space of regular (cusp) forms on Γ of dimension $-r$	97
$\tilde{c}$		45
	conjugate side	37
	cycle (ordinary, elliptic, parabolic, accidental)	39,39,42,39
$D(f)$	divisor of f	128
$D(d\Omega)$	divisor of $d\Omega$	130
$d(a, b)$	H-distance between a and b	24
$d\Omega_m, d\Omega$	differential of weight m, weight 1	127
$\|d\tau\|/y$	differential of H-length	23
deg D	degree of the divisor D	87, 94
dim D	dimension of space of multiples of D	128
	differential of H-area	25
	differential of the first kind	130
	discontinuous group	11
	discrete group	13
E	real axis	
$e(u)$	$\exp(2\pi iu)$	
	elliptic (parabolic)sector	41, 44
	elliptic, hyperbolic, parabolic linear transformation	7
$\eta(q,\varphi)$	order of φ at q	126
$F\|S$ or F_S	S-transform of F	73
$\hat{f}(t), \hat{F}(t)$		82-84, 92
	free side	33
	Fuchsian group of first (second) kind	21
	fundamental region (set)	22
$G_{-r}(\tau, \nu)$	Poincaré series with parameter ν	68
$(G/H), (H\backslash G)$	system of coset representatives	67,ftn.
g	genus	124
$\Gamma x = [x]$	orbit of x	11, 117
$\Gamma(1)$	modular group	12

Symbol	*Description*	*Page*
$\Gamma(n)$	principal congruence subgroup of level n	12
$\Gamma^0(p), \Gamma_0(p)$		62
Γ_x	stabilizer of x	14
$\Gamma \backslash H^+$		117
$\{\Gamma, -r\}$	space of automorphic forms on Γ of dimension $-$r	76
	Γ-equivalence	11
	Gauss-Bonnet formula	48
H	upper half-plane $\{\operatorname{Im} z > 0\}$	
	H-disk (circle)	25
	H-line (point, plane)	23
	horocyclic	21
$\mathbf{I}(T)$	isometric circle of T	57
$K(S)$	field of meromorphic functions on S	131
$\mathbf{K}(\Gamma)$	field of automorphic functions on Γ	131
L	limit set of a group	10
$l = l(x)$		51
	linear transformation	3
	multiple of a divisor	128, 130
	multiplier of a linear transformation	6
N	normal polygon	28
$\|N\|$	hyperbolic area of the normal polygon N	51
$N(f)$	valence of f	84
$n(\tau, f)$, $n(\tau, F)$	order of f, F at τ	82-84, 92
$\mathcal{O}$	set of ordinary points of a group	10
Ω_Q		9
Ω_R		8
$\bar{\Omega}_C, \Omega_C$		3,4
	parametric disk	126
Φ_x		119
S_x, S^*_x		115,119
σ		118
	Scalar Product Formula	100
	side	31
$\{T_1, T_2, \ldots\}$	group generated by T_1, T_2,	47
$t = \Phi\ (q)$	local variable	117
$\tau_x(z)$		120
U	unit disk $\{\|z\| < 1\}$	
U_x		119
	vertex (ordinary, real, elliptic, parabolic, accidental)	32,34,39,42,39
Z	canonical class	130
Z	complex sphere	

Index

A CATALOG OF SELECTED

DOVER BOOKS

IN SCIENCE AND MATHEMATICS

Mathematics–Bestsellers

HANDBOOK OF MATHEMATICAL FUNCTIONS: with Formulas, Graphs, and Mathematical Tables, Edited by Milton Abramowitz and Irene A. Stegun. A classic resource for working with special functions, standard trig, and exponential logarithmic definitions and extensions, it features 29 sets of tables, some to as high as 20 places. 1046pp. 8 x 10 1/2. 0-486-61272-4

ABSTRACT AND CONCRETE CATEGORIES: The Joy of Cats, Jiri Adamek, Horst Herrlich, and George E. Strecker. This up-to-date introductory treatment employs category theory to explore the theory of structures. Its unique approach stresses concrete categories and presents a systematic view of factorization structures. Numerous examples. 1990 edition, updated 2004. 528pp. 6 1/8 x 9 1/4. 0-486-46934-4

MATHEMATICS: Its Content, Methods and Meaning, A. D. Aleksandrov, A. N. Kolmogorov, and M. A. Lavrent'ev. Major survey offers comprehensive, coherent discussions of analytic geometry, algebra, differential equations, calculus of variations, functions of a complex variable, prime numbers, linear and non-Euclidean geometry, topology, functional analysis, more. 1963 edition. 1120pp. 5 3/8 x 8 1/2. 0-486-40916-3

INTRODUCTION TO VECTORS AND TENSORS: Second Edition--Two Volumes Bound as One, Ray M. Bowen and C.-C. Wang. Convenient single-volume compilation of two texts offers both introduction and in-depth survey. Geared toward engineering and science students rather than mathematicians, it focuses on physics and engineering applications. 1976 edition. 560pp. 6 1/2 x 9 1/4. 0-486-46914-X

AN INTRODUCTION TO ORTHOGONAL POLYNOMIALS, Theodore S. Chihara. Concise introduction covers general elementary theory, including the representation theorem and distribution functions, continued fractions and chain sequences, the recurrence formula, special functions, and some specific systems. 1978 edition. 272pp. 5 3/8 x 8 1/2. 0-486-47929-3

ADVANCED MATHEMATICS FOR ENGINEERS AND SCIENTISTS, Paul DuChateau. This primary text and supplemental reference focuses on linear algebra, calculus, and ordinary differential equations. Additional topics include partial differential equations and approximation methods. Includes solved problems. 1992 edition. 400pp. 7 1/2 x 9 1/4. 0-486-47930-7

PARTIAL DIFFERENTIAL EQUATIONS FOR SCIENTISTS AND ENGINEERS, Stanley J. Farlow. Practical text shows how to formulate and solve partial differential equations. Coverage of diffusion-type problems, hyperbolic-type problems, elliptic-type problems, numerical and approximate methods. Solution guide available upon request. 1982 edition. 414pp. 6 1/8 x 9 1/4. 0-486-67620-X

VARIATIONAL PRINCIPLES AND FREE-BOUNDARY PROBLEMS, Avner Friedman. Advanced graduate-level text examines variational methods in partial differential equations and illustrates their applications to free-boundary problems. Features detailed statements of standard theory of elliptic and parabolic operators. 1982 edition. 720pp. 6 1/8 x 9 1/4. 0-486-47853-X

LINEAR ANALYSIS AND REPRESENTATION THEORY, Steven A. Gaal. Unified treatment covers topics from the theory of operators and operator algebras on Hilbert spaces; integration and representation theory for topological groups; and the theory of Lie algebras, Lie groups, and transform groups. 1973 edition. 704pp. 6 1/8 x 9 1/4. 0-486-47851-3

Mathematics-Geometry and Topology

PROBLEMS AND SOLUTIONS IN EUCLIDEAN GEOMETRY, M. N. Aref and William Wernick. Based on classical principles, this book is intended for a second course in Euclidean geometry and can be used as a refresher. More than 200 problems include hints and solutions. 1968 edition. 272pp. 5 3/8 x 8 1/2. 0-486-47720-7

TOPOLOGY OF 3-MANIFOLDS AND RELATED TOPICS, Edited by M. K. Fort, Jr. With a New Introduction by Daniel Silver. Summaries and full reports from a 1961 conference discuss decompositions and subsets of 3-space; n-manifolds; knot theory; the Poincaré conjecture; and periodic maps and isotopies. Familiarity with algebraic topology required. 1962 edition. 272pp. 6 1/8 x 9 1/4. 0-486-47753-3

POINT SET TOPOLOGY, Steven A. Gaal. Suitable for a complete course in topology, this text also functions as a self-contained treatment for independent study. Additional enrichment materials make it equally valuable as a reference. 1964 edition. 336pp. 5 3/8 x 8 1/2. 0-486-47222-1

INVITATION TO GEOMETRY, Z. A. Melzak. Intended for students of many different backgrounds with only a modest knowledge of mathematics, this text features self-contained chapters that can be adapted to several types of geometry courses. 1983 edition. 240pp. 5 3/8 x 8 1/2. 0-486-46626-4

TOPOLOGY AND GEOMETRY FOR PHYSICISTS, Charles Nash and Siddhartha Sen. Written by physicists for physics students, this text assumes no detailed background in topology or geometry. Topics include differential forms, homotopy, homology, cohomology, fiber bundles, connection and covariant derivatives, and Morse theory. 1983 edition. 320pp. 5 3/8 x 8 1/2. 0-486-47852-1

BEYOND GEOMETRY: Classic Papers from Riemann to Einstein, Edited with an Introduction and Notes by Peter Pesic. This is the only English-language collection of these 8 accessible essays. They trace seminal ideas about the foundations of geometry that led to Einstein's general theory of relativity. 224pp. 6 1/8 x 9 1/4. 0-486-45350-2

GEOMETRY FROM EUCLID TO KNOTS, Saul Stahl. This text provides a historical perspective on plane geometry and covers non-neutral Euclidean geometry, circles and regular polygons, projective geometry, symmetries, inversions, informal topology, and more. Includes 1,000 practice problems. Solutions available. 2003 edition. 480pp. 6 1/8 x 9 1/4. 0-486-47459-3

TOPOLOGICAL VECTOR SPACES, DISTRIBUTIONS AND KERNELS, François Trèves. Extending beyond the boundaries of Hilbert and Banach space theory, this text focuses on key aspects of functional analysis, particularly in regard to solving partial differential equations. 1967 edition. 592pp. 5 3/8 x 8 1/2. 0-486-45352-9

INTRODUCTION TO PROJECTIVE GEOMETRY, C. R. Wylie, Jr. This introductory volume offers strong reinforcement for its teachings, with detailed examples and numerous theorems, proofs, and exercises, plus complete answers to all odd-numbered end-of-chapter problems. 1970 edition. 576pp. 6 1/8 x 9 1/4. 0-486-46895-X

FOUNDATIONS OF GEOMETRY, C. R. Wylie, Jr. Geared toward students preparing to teach high school mathematics, this text explores the principles of Euclidean and non-Euclidean geometry and covers both generalities and specifics of the axiomatic method. 1964 edition. 352pp. 6 x 9. 0-486-47214-0